STUDIES IN THE ARCADIAN LIBRARY

NO. 4

SCIENTIFIC EXPEDITIONS TO THE ARAB WORLD (1761–1881)

by

J.M.I. Klaver

THE ARCADIAN LIBRARY
in association with
OXFORD UNIVERSITY PRESS

OPPOSITE TITLE PAGE
Jean-Constantin Protain, engraving of the garden of the Institut d'Égypte. *Description de l'Égypte*, Paris 1809, *État Moderne*, vol. 1, Villes et sites de Haute et Basse Egypte, plate 56

The Arcadian Library in association with Oxford University Press

The Arcadian Library, London, is an affiliate of
The Arcadian Group S.A., Case Postale 200, 1211 Geneva 12, Switzerland

The Arcadian Library is a private, family library specializing in the historic influences of the Levant upon Europe. Its aim is to promote this cultural transfer through exhibition and through publication

Oxford University Press, Great Clarendon Street, Oxford OX2 6DP

Oxford University Press is a department of the University of Oxford.
It furthers the University's objective of excellence in research, scholarship, and education by publishing worldwide in Oxford New York Auckland Cape Town Dar es Salaam Hong Kong Karachi Kuala Lumpur Madrid Melbourne Mexico City Nairobi New Delhi Shanghai Taipei Toronto
With offices in Argentina Austria Brazil Chile Czech Republic France Greece Guatemala Hungary Italy Japan Poland Portugal Singapore South Korea Switzerland Thailand Turkey Ukraine Vietnam

Oxford is a registered trade mark of Oxford University Press
in the UK and in certain other countries

Published in the United States by Oxford University Press Inc., New York

First published 2009

British Library Cataloguing in Publication Data (data available)

Library of Congress Cataloguing in Publication Data (data available)

ISBN 978-0-19-956889-5

Project management by Robert Jones, Series editor Alastair Hamilton
Copy-editor Bernard Dod, Designed and typeset by Phil Cleaver and Emily Berry of etal-design.com ltd
Photography at The Arcadian Library: John Stone and Karem Ibrahim
Printed in Great Britain by Butler Tanner & Dennis, on Stow Book Wove, free from chlorine (TCF) and OBA, and resistant to ageing in accordance with ISO 97006
Typeset in Monotype Arcadian Bembo

TABLE OF CONTENTS

PREFACE

When the French army advanced up the Nile in 1799 and suddenly came upon the panorama of the ancient city of Thebes, an 'émotion électrique' ran through the soldiers, and, spontaneously halting, they clapped their hands. It was a moment of enthusiasm in a prolonged and futile military campaign which was full of hardship and disease. But it was with inspiring reports and many drawings of ancient Egypt, often made in haste, that the French returned to Cairo. A historian commented that if any conquest had been made, the conquest was theirs, and it was never lost. The sentiment is emblematic for the scientific work of the expeditions described in this study.

Hand-coloured engraving of juvenile Spotted Eagle. *Description de l'Égypte*, Paris 1808, [proof copy of 1809 edition], *Histoire naturelle*, vol. 1, Zoologie-Oiseaux, plate 2
The Arcadian Library, 09142

This book is about the great scientific expeditions which set out to catalogue and study the flora and fauna of the Near and Middle East. In 1761 a Danish expedition visited Egypt, the Red Sea, and the Southern Arabian peninsula, while at the turn of the century the French invasion of Egypt led to a thorough investigation of the natural history of that country. From 1820 to 1825 a Prussian expedition followed in the footsteps of the Danes and the French. At the end of the century German and British scientific expeditions were set up to study the endemic flora and fauna of the island of Soqotra. They were all expensive enterprises, state-sponsored and performed in the modern framework of Linnaean nomenclature.

Of course, over the centuries merchants, diplomats, missionaries and travellers had visited Egypt and the Arabian peninsula and made notes of the unknown plants and animals they encountered. But the study of the natural history of the Middle East remained only a part of a larger description of their travel experiences, and very little systematic research was carried out. On the other hand, the post-Linnaean Danish, French, German and English expeditions sent out trained botanists and zoologists with the specific object of collecting and mapping out all the species of the area. The efforts of these naturalists are scientifically important: many of the plants and animals they collected constitute the type specimens of species we recognize today.

What the naturalists on the expeditions studied in this book have in common is that they were all highly qualified young men who hoped to make their name in the European scientific community by their work on the flora and fauna of the Near and Middle East. Although they were not blind to the dangers of spending years of study in far-away countries, they all left with high hopes of achieving important results and they were not without a spirit of adventure. A strong sense of disillusionment, however, accompanies their final scientific achievements. Many of the naturalists succumbed to disease and never returned to Europe. And many of those who did return lost interest in working out the results of the expeditions and left their discoveries for others to describe and publish. This book follows the personal vicissitudes of the naturalists on the expeditions, analysing their feelings as their research progressed, and assesses the scientific importance of their work.

In writing this book I have received much support. I am particularly grateful to the chairman of The Arcadian Library for giving me access to the Library's fine collection and for his dedication to publishing research on the historical influences of the Levant upon Europe. I am also grateful for the help and encouragement of Alastair Hamilton and Robert Jones. Their suggestions and comments on the manuscript have been much appreciated. Thanks are also due to Professor Colin Groves, Professor Harold Cook and Professor Janet Browne for reading my manuscript and for their valuable observations. A number of institutions and individuals have assisted me in my research. I would like to thank former and present staff of The Arcadian Library: Alice Bailey and Karem Ibrahim for their help at the library during the early stages of my research, Daniela Reimers for indefatigably looking up bibliographic information, and Andrea Meyer Ludowisy for assistance during the last stages of publication. The reproduction department of the British Library kindly produced an electronic copy of Balfour's *Botany of Socotra* for me, and Giuseppe Ambrogi of Urbino University Library was indispensable in tracing and borrowing publications from Danish and German libraries. I would also like to thank Dr Maurits van den Boogert for providing me with photocopies of Niebuhr's letters. Special thanks are due to Lynda Brooks at the Linnean Society for her hospitality at the society's library and for allowing the reproduction of rare botanical and zoological plates. I am also in debt to Simone Trieder who generously shared the results of her research on the Riebeck family with me. Amongst those who patiently answered my questions in different fields I would like to thank Professor Ib Friis, Professor Michael Dallapiazza, Dr Antonella Negri and Professor Gioia Zaganelli. I must also thank John Stone for photography of the plates and Phil Cleaver at etal-design for the superb layout of the volume. Special thanks are due to Bernard Dod whose professional expertise helped to see the book through the editorial and production process. Finally I would like to thank my wife Anna for her support in writing this book, and my father, to whom this volume is dedicated, for instilling a lively interest in nature into me.

EUROPE AND THE SCIENTIFIC EXPLORATION OF THE NEAR AND MIDDLE EAST

During the 150 years following Linnaeus' groundbreaking publications in the mid-1750s, a number of full-scale scientific expeditions to the Arab countries around the Red Sea were undertaken with the purpose of describing and classifying unknown species of animals and plants. Denmark was the first nation to organize such an expedition, sending a group of five scholars and scientists to Egypt and the Yemen in 1760. Forty years later Napoleon took a number of naturalists with him on his Egyptian campaign, founding a counterpart of the Institut de France in Cairo. In the 1820s the University of Berlin financed yet another scientific expedition to explore the flora and fauna of the Near and Middle East. As far as the study of nature was concerned, these expeditions must be seen against the late eighteenth- and nineteenth-century urge to stock the natural history collections of European museums with specimens of plant and animal species from all over the world, a phenomenon caused not least by questions of national prestige.

Although the Danish, French and German expeditions stand out as huge and expensive projects of research in the natural history of the Middle East, many other European travellers had visited the area, mapped out its geography, studied its plants and described its animals and rocks. The most famous work from antiquity is no doubt Herodotus' *History*. In the passages on Egypt in the second book (also known as *Euterpe*), written in the mid-fifth century B.C., Herodotus intersperses factual information about local customs, religion, geography, and the climate with occasional tales of wonder. Although he does not dedicate much space to natural history, there are some descriptions of animals. For example, he mentions the crocodile's scaly hide, its strong moving upper jaw, its near-blindness when in the water, and the absence of a tongue, and he describes the habit of the *trochilus* in liberating the reptile from its leeches.[1] He also reports on 'river-horses' (hippopotamuses), otters and eels. He mentions the phoenix, but admits that he himself did not see the animal. Finally, in an interesting passage, he refers to winged serpents and the role of the ibis as the guardian of Egypt:

there is a place in Arabia to which I went to learn about the winged serpents. When I came thither, I saw innumerable bones and backbones of serpents; many heaps of backbones there were, great and small and smaller still. This place, where lay the backbones scattered, is where a narrow mountain pass opens into a great plain of Egypt. Winged serpents are said to fly at the beginning of spring, from Arabia, making for Egypt; but the ibis birds encounter the invaders in this pass and kill them.[2]

Herodotus' book was seen by many later naturalists as a starting point, and the

1. Herodotus, *History*, transl. by A. D. Godley, The Loeb Classical Library, 4 vols., London 1960, II.68, pp. 355–7.
2. Herodotus, *History*, II.75, pp. 361–3.

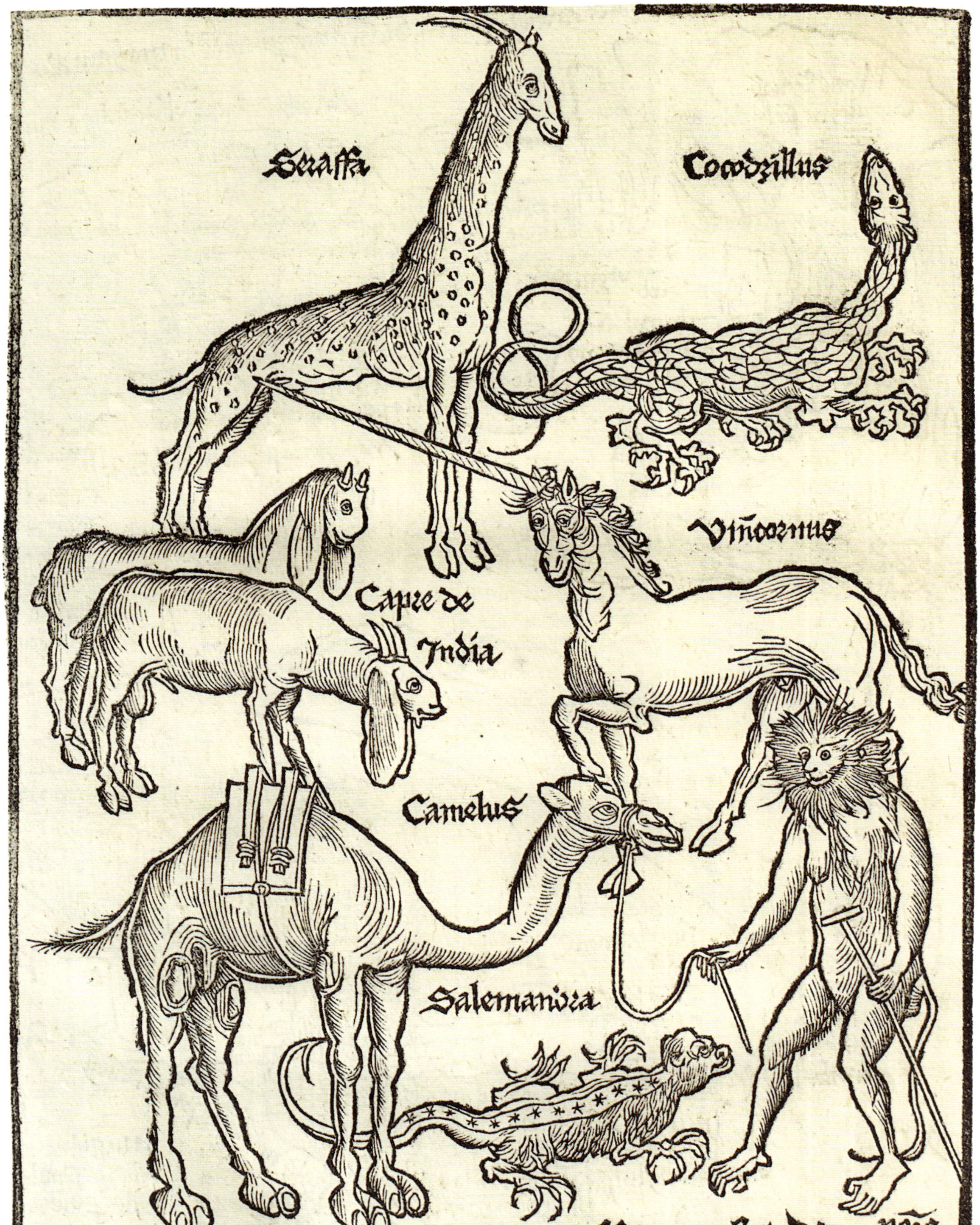

Seraffa
Cocodrillus
Vñicornus
Capre de
India
Camelus
Salemandra
Non constat de
noĩe

French zoologists Geoffroy Saint-Hilaire and Savigny analysed Herodotus' descriptions in their nineteenth-century studies of the crocodile and the ibis.

The standard natural history work of classical antiquity was the huge compilation by the encyclopedist Pliny the Elder (23–79 A.D.) who described in 37 books all that was known at the time about the botany, zoology, mineralogy, geography and ethnology of Europe and the Near East. But, unlike Herodotus, Pliny did not travel widely—he never visited the Near East—and the greater part of his descriptions were based on second-hand observations. Most of his zoological data, for example, derived from Aristotle, while many of his botanical observations were based on the works of Theophrastus. Although Pliny's *Natural History* included legendary information gleaned from folklore (such as his description of the phoenix and his belief that chameleons subsisted on air), it did make a genuine contribution to natural science, especially in the books on botany, and he remained a virtually uncontested authority throughout the Middle Ages until the end of the fifteenth century.

The Dean of Mainz, Bernhard von Breydenbach, was one of a group of German pilgrims who travelled to the Holy Land in the late Middle Ages. He reached Jerusalem in July 1483, visited St Catharine's, and then proceeded to Cairo and Alexandria. After his return to Europe he published an account of his experiences, interspersed with the descriptions of earlier travellers. His *Peregrinatio in Terram Sanctam* (1486) was illustrated by the Dutch artist and engraver Erhard Reuwich, who accompanied him on his journey. The volume was highly successful in its time—no fewer than twelve editions appeared between 1486 and 1522—and it is generally regarded as the first illustrated travel book. Although the text is of no interest for the natural history of the Near East, one illustration towards the end of the volume is. It shows a group of animals which Breydenbach and Reuwich had seen, including a dromedary, two Indian goats, a giraffe, a crocodile and a salamander. It also featured a drawing of a unicorn and a hominoid baboon. This is probably the first occasion on which a giraffe and a dromedary were represented in an illustration, but though the latter is tolerably realistic, the giraffe has the horns of a gazelle and there is something very wrong with its tail. Although the caption underneath the illustration states that the animals were truthfully depicted,[3] it seems obvious that Reuwich's memory failed him when he drew them at a later stage.

Bernhard von Breydenbach, woodblock print of giraffe, crocodile, Indian goats, unicorn, dromedary, salamander and hominoid.
Dis Buch ist innhaltend die heiligen reysen gein Jherusalem zu dem heiligen grab und furbaß zu der hochgelobten jungfrowen und merteryn sant katheryn, Speyer 1505 edn.
The Arcadian Library, 08858

The first and most influential Renaissance text on the natural history of the Near East was by the French naturalist Pierre Belon du Mans. Belon explored the Ottoman countries of the eastern Mediterranean between 1546 and 1549, using as a base the residence of Gabriel d'Aramon, the French ambassador in Istanbul. He explored part of Greece, Turkey and Palestine, and, during the late summer and early autumn of 1547, visited Egypt, which had been seized by the Turks in 1517. His travel account, *Observations des plusieurs singularitez et choses memorables trouvées en Grèce, Asie, Iudée, Égypte, Arabie, et autres pays estranges* (1553), contains numerous descriptions of animals and plants based on first-hand observation. With his *Observations* and the woodcuts in *Portraits d'oyseaux, animaux, serpens, herbes, arbres, hommes et femmes, d'Arabie et Égypte* (1557) Belon charted many species of plants and animals from the Near East that were unknown to Europeans. He drew up lists of the

3. '[H]ec animalia sunt veraciter depicta sicut vidimus in terra sancta,' *Peregrinatio in Terram Sanctam*, caption to unnumbered plate in the original Latin edition.

Portraict de la Giraffe.

quand elle court, les deux piedz de deuant vont ensemble. Elle se couche le ventre contre terre, & a vne durté à la poictrine & aux cuisses comme vn chameau. Elle ne sçauroit paistre en terre estant debout sans eslargir grandement les iambes de deuant, encore est ce auec grande difficulté. Parquoy il est aisé à croire qu'elle ne vit aux champs sinon des branches des arbres, ayant le col ainsi long, tellement qu'elle pourroit arriuer de la teste à la haulteur d'vne demie picque. Et l'ayant faict retirer au naturel, i'en ay bien voulu icy mettre le portraict.

D'VN MOVLT BEAV PETIT BOEVF d'Aphrique, que les anciens Grecs nommerent Bubalus. Chapitre L.

LE plaisir qu'vn homme curieux peult receuoir de rencontrer vn animal estrange & singulier, est de luy trouuer quant & quant son nom ancien, pour le sçauoir exprimer: car celuy qui a quelque chose à descrire, sans la nommer de nom propre, me semble faire coruée d'en prendre la peine. Parquoy m'estant trouué à veoir vn petit bœuf d'Affrique, trappe & ramassé, gras, poly, de petit corsage, bien formé, soudainement me tumba en la memoire que c'estoit celuy que les Grecs auoient anciennement nommé Bubalos: mais fish he saw around Alexandria, inspected those caught by the fishermen of the Nile, and carefully observed the birds during his voyage up to Cairo. He went out to catch snakes and had them stuffed. He described numerous plants, especially trees, and dwelt on their medicinal properties. He was also the first scholar to produce a reliable picture of a giraffe, drawn after a specimen in a menagerie in Cairo.[4] His description of what he called a red lily makes him the first westerner to describe the tulip. Despite his inclusion of some fantastic entities, such as the phoenix and the flying serpent, his work was of genuine scientific value, based on direct observation. He did away, for example, with Pliny's belief that chameleons lived on air by dissecting one. He found that, although they could indeed live a long time without food, they preyed on small animals like grasshoppers, snails and caterpillars, and needed to drink water to survive.[5] Of further interest in the history of science is the fact that Belon draws attention to the homologous structure of the skeleton of a bird and that of a human.[6] It makes him an early forerunner of Geoffroy Saint-Hilaire, who joined Napoleon's Egyptian campaign two and half centuries later and happened to advance his own theory of homologies while studying the fish of the Nile.

Whereas Belon was important for collecting reliable information about the animals and plants of the Levant, the Habsburg diplomat Augier Ghislain de Busbecq, ambassador to the Sultan, brought live specimens of camels, horses, Angora goats and a mongoose to Europe in 1562. Busbecq was also an enthusiastic herbalist and he returned from his travels with the lilac, the sedge, the horse chestnut and new species of irises, ranunculi and anemones. He reported on cultivars of the tulip, which he called 'Tulipa Turcarum', and some historians even attribute to him its introduction into Europe. In Turkey he obtained the most ancient extant codex of Dioscorides' *De materia medica* . This copy of the famous herbal book is known as the Vienna Dioscorides and is now in the Österreichische Nationalbibliothek.

In the following decade the German traveller and physician Leonhart Rauwolff visited the Levant. Former botany student of Rondelet in Montpellier, Rauwolff was sent out to find profitable plants and drugs for a firm of merchants. In 1573 he first travelled to Tripoli and Aleppo, and from there to Baghdad, and then back via Jerusalem. After his return to Augsburg in 1576 he carefully sorted his botanical collection from the Near East in a herbarium (his 'Viertes Kreutterbuech') and, in his travel account *Aigentliche Beschreibung der Raiß ... inn die Morgenländer* (1582–3), he published excellent descriptions and fine woodcuts of the plants he had found. The rare, seamless transitions from travel narrative to scientific description would later only be matched by Niebuhr's account of the Danish expedition of 1761–7. The following is an example of a late seventeenth-century translation of Rauwolff's text. It was published in John Ray's *Collection of Curious Travels and Voyages* (1693):

Pierre Belon, hand-coloured woodblock print of a giraffe. *Les Observations de plusieurs singularitez et choses mémorables trouvées en Grèce, Asie, Iudée, Égypte, Arabie et autres pays estranges*, Paris 1554 edn., p. 119v
The Arcadian Library, 08185

NEXT OPENING
Pierre Belon, hand-coloured woodblock print of a crocodile. *Les observations de plusieurs singularitez et choses mémorables trouvées en Grèce, Asie, Iudée, Égypte, Arabie et autres pays estranges*, Paris 1554 edn., pp. 102v–103r
The Arcadian Library, 08185

4. Pierre Belon du Mans, *Portraits d'oyseaux, animaux, serpens, herbes, arbres, hommes et femmes, d'Arabie et Égypte*, Paris 1557, p. 105. It was not uncommon to see giraffes in a Cairo menagerie. Nicholas Warner in *The True Description of Cairo* says the embassy from Sultan Qaytbay that visited Florence in 1487 brought a giraffe and several lions as gifts (*The True Description of Cairo: A Sixteenth-Century Venetian View*, 2 vols., Oxford 2006, I.139).

5. Pierre Belon du Mans, *Les observations de plusieurs singularitez et choses mémorables, trouvées en Grèce, Asie, Iudée, Égypte, Arabie, et autres pays estranges, redigées en trois livres*, Paris 1553, pp. 125–6.

6. Belon, *Portraits*, pp. 3–4.

DESCRIPTION DE PLVSIEVRS OISEAVX, & autres animaulx obseruez le long du Nil. Chapitre XXXII.

*L*E pays d'Egypte estant si tiede l'hyuer & palustre, nourrit plusieurs oiseaux de riuiere, & entre autres celuy que les Grecs & Aristote ont nommé Crex. Ie l'ay recogneu à sa voix, car il est criart, & comme le Vanneau dit AEx, tout ainsi cestuy cy en voulant prononcé Crex Crex. Ie le descripui, lors comme s'ensuit. L'oiseau nommé Crex, est de corsage entre les Corlis & le Cheuallier, ayant aussi le bec & les iambes entre les deux, ses iambes, cuisses & pieds sont noirs, comme aussi est sa teste: Mais le dessus du col, la poictrine & espaules sont blanches, le dessus du corps tient du cendré, ayant vne ligne blanche de trauers en chasque ælle. Il prend sa mangeaille en terre, & en l'air, à la maniere du Vanneau, que les anciens Grecs ont nommé AEx, & faict ainsi grand bruit des ælles en volant. Ie croy qu'il n'est point veu entre les oiseaux cognus de noms François, combien que i'eusse pensé auparauant que la Barge estoit Crex, entant que Herodote l'a comparé en grandeur à l'vne des especes de l'oiseau nommé Ibis. I'auoye au parauant escript cest Ibis noir, pensant qu'il fut Hæmatopus: mais ayant depuis obserué ses meurs, i'ay arresté que ce n'est Hæmatopus, ains Ibis noir, duquel Herodote premierement a faict mention, puis apres luy Aristote. Il est de corpulence d'vn Corlis, ou quelque peu moindre, totalement noir, ayant la teste d'vn Cormarant, le bec contre la teste plus gros que le poulce, mais poinctu par le bout & voulté, & quelque peu courbe, & tout rouge, cõme aussi les cuisses & les iambes. Il est tout ainsi hault eniambé comme vn Butor, que Pline a nommé Bos taurus, & Aristote Ardea stellaris, & a le col ainsi long que vne Aigrette, en sorte que quand ie vei ledict Ibis noir la premiere fois, il me sembla en habitude & contenance à vn Butor. Les Egyptiens, Mores ou Arabes, sont plus superstitieux & ceremonieux en leur religion que ne sont les Turcs: & ia soit que tous deux soient d'vne mesme loy, tendants à Mahomet, & subiectz au grand Turc, qui les a vaincus en bataille, toutesfois les Turcs estiment quelque sainctetè es Arabes plus qu'en eulx mesmes. La raison est que l'Alcoran fut escript en Arabe, qui a depuis esté translaté en Turquois, & aussi que les plus doctes Turcs ne font pas profession du langage Turquoys, mais de l'Arabic. Leurs characteres sont vne mesme chose, toutesfois la langue est diuerse. Aussi les Turcs n'ont point de lettres qui ne

Crex. Aex. Vanneau. Corlis. Cheuallier. Vanneau. Barge. Ibis noir. Hæmatopus. Cormarãt. Butor. Aigrette. Arabes ceremonieux. Alcoran.

ne soient venues des Arabes. Quand nous descendions du batteau aux riuages du Nil pour entrer es villages, nous entendions les Mores chanter en leurs mosquées, c'est à dire Eglises, qui se respondent les vns aux autres de voix alternatiues, à la maniere des prestres Latins, faisant quasi mesmes accens, & mesmes pausées, comme font ceulx qui chantent les Pseaumes en Latin : qui est chose qu'on ne faict point entre les Turcs, qui ont dur langage & rude à la comparaison de l'Arabe, qui est moult aisé à toutes choses qu'on veult mettre en rythme. Ausi l'Alcoran est escript en verses de rythme. En approchant du Caire, à quatre lieues au dessoubs de la ville, nous vismes l'endroict ou le canal du Nil se depart en deux rameaux : desquels l'vn descendant à gauche, & va passer à Rosette, qui est Ostium Canopicum, d'ou nous venions. L'autre descend à dextre, & se rend en Damiate, ou est Ostium Pelusiacum. Par cela nous pouuons asseurer que le Nil n'a que deux principales grandes bouches nauigables pour grands vaisseaux, ou pour le plus en a trois grandes en tout. Ie ne dy pas qu'il n'ait beaucoup de petis ruisselets, mais il n'a que ces deux principaux nauigables. Il peult bien estre que quelques vns sont nauigables en certains endroicts au temps de l'inondation : mais en autre temps ce sont petis ruisseaux qu'on passe à gué au riuage de la mer, comme nous auons faict quand nous auons passé le petit canal entre Alexandrie & Rosette. Le vent nous continua iusques au Caire, ou nostre nauigation finit. Descendismes à vn grand village nommé Boulac, qui est du tenant du Caire, situé au riuage du Nil. Auant me deporter de parler du Nil, ie diray premierement de quelques bestes qu'on a accoustumé d'y trouuer, & entre autres du Crocodile dont cy est le portraict.

Langue Turcoise

Le departement du Nil.

Ostiú Canopicum.

Ostiú Pelusiacum.

Damiete.

Deux bouches du Nil.

Boulac.

Portraict du Crocodile poisson du Nil.

OPPOSITE
Leonhart Rauwolff, opening of the third part of *Aigentliche beschreibung der Raiß so er vor dieser zeit gegen Auffgang inn die Morgenländer fürnemlich Syriam, Iudaeam, Arabiam, Mesopotamiam, Babyloniam, Assyriam, Armeniam etc. nicht ohne geringe mühe vnd grosse gefahr selbs volbracht*, Lauingen 1583, p. 297
The Arcadian Library, 12796

NEXT OPENING
Leonhart Rauwolff, woodblock prints of aubergines.
Aigentliche beschreibung der Raiß so er vor dieser zeit gegen Auffgang inn die Morgenländer fürnemlich Syriam, Iudaeam, Arabiam, Mesopotamiam, Babyloniam, Assyriam, Armeniam etc. nicht ohne geringe mühe vnd grosse gefahr selbs volbracht, Lauingen, 1583, plate 73
The Arcadian Library, 12796

FOLLOWING TWO OPENINGS
David Loggan, engraved portrait of John Chardin and title page.
John Chardin, *Journal du Voyage ... en Perse et aux Indes Orientales, par la Mer Noire et par la Colchide. Première partie, qui contient le voyage de Paris à Ispahan*, London 1686
The Arcadian Library, 11218

Engraving and description of a rhinoceros in the royal stables in the Persian capital Ispahan.
John Chardin, *Voyages ... en Perse, et autres lieux de l'Orient*, Amsterdam 1711 edn., III.45
The Arcadian Library, 11221

I saw hereabouts two thorny shrubs, one whereof was, with it's [sic] red colour'd grapes (setting the leaves aside as far as I remember) very like unto our Oxyacantha. *The other, which was full of thorns, and had small red purple colour'd flowers, was like unto the firsy kind of* Scorpius *of* Carolus Clusius. *Among the bushes I saw the Scorzonera with yellow flowers, and also a pretty sort of Tulips with yellow stripes [...]*

The mountain is very high, so that it may be seen in Cyprus *about 200* Italian *miles off; wherefore the day breaketh later at* Tripoli, *and not until the morning sun appeareth before it: And moreover you find there snow all summer long, which they bring down from the mountains, into the Batzars or Exchanges to sell, to cool their drink with it.*[7]

After Rauwolff's death the Dutch humanist Isaac Vossius came across a copy of his herbal, now preserved in the library of Leiden University. Rauwolff's herbarium and observations were of such outstanding quality that many of them were used by the eighteenth-century Dutch botanist Jan Fredrick Gronovius to produce Linnaean descriptions of them in his *Flora orientalis* of 1755.

In 1580 yet another physician and botanist travelled to the Levant. As personal doctor to the Venetian consul in Cairo, Prospero Alpini spent three years in Egypt. During his stay he studied the local flora around Cairo as well as the various plants he found growing in its gardens. The results of these studies were published in his *De plantis Aegypti* (1592), which features 57 descriptions and 49 illustrations of plants that were mainly unknown in the West. In his *De medicina Aegyptorum* he brought the coffee bush, the banana and the boabab to European attention. In 1591 he also published a dialogue on the balsam tree of the ancients, of which he had brought a specimen to Venice. He maintained that the tree was not of Egyptian, but Arabian origin, and his description made it a controversial species for almost two centuries, until it was finally found by Pehr Forsskål near Sana'a in the Yemen. In his description of the date palm Alpini noted that fertilization of plants was a sexual process. This was later to become the foundation of the Linnaean system of taxonomy. In 1593 he was appointed professor of botany at the University of Padua.

Often as a result of the increasing diplomatic relations with the Ottoman Empire numerous European travellers continued to visit, and describe, the countries on the eastern Mediterranean. Yet relatively little of note was done in the field of botanical or zoological exploration of the Levant during the seventeenth century. Early in the century the Portuguese Jesuit Jeronimo Lobo travelled to Ethiopia, one century after Francisco Alvares, but from his accounts only the merest mention of a handful of animals and plants can be gleaned. The German missionary Johann Michael Wansleben travelled to Tripoli, Damascus, Aleppo and Egypt in 1672–3 (but not to Ethiopia which was supposed to be his destination). Although he was generally a keen observer, he had little interest in natural history and reported vague and baseless explanations of the mating rituals of the crocodile. In 1675 the diplomat Francis Vernon travelled to the Ottoman Empire with George Wheeler and the French archaeologist Jacob Spon. Vernon went no further than Athens. Wheeler, on the other hand, continued to Istanbul on his own and drew up a good list (with short descriptions in Latin) of the plants he found in both Greece and Turkey.

7. John Ray, *Collection of Curious Travels and Voyages*, London 1738, p. 192.

Der Dritte Thail

der Rayß/ Doctor Leonhart Rauwolffs/ in die Morgenlender: inn welchem fürnemlich wirt begriffen die beschreibung der statt Jerusalem/ jrer vmbligenden Landtschafft/ vnd etlicher heiliger örter/ so zů diser zeit noch jnner vnd ausserhalb zů sehen/ vnd einem jeden Bilgram gewisen werden.

Darneben erzelt werden/ der Türcken erstlich/ vnnd hernacher auch etlicher Christen leben/ Glauben vnnd jrrthumb/ sonderlich aber deren/ so dem Türckischen Kayser vnderworffen/ vnd sich im Tempel deß Berges Caluariæ verhalten.

Melantzana, von Arabern Melongena vñ Bedengian genannt/ deren dreyerley bey jnen zůfinden/ die in jren farben zů vnderschiden/ gelb/ äschen vnnd leibfarbe.

Nū: 73.

Schwartze

Schwartze Melantzana, die vmb ein zimlichs grösser/ deren zweyerley/ langlecht vnd runde seind/ den Moren vnder dem namen Bathleschain bekannt.

Nū: 73.

Dom=

JOHANNES CHARDIN MILES.

Natus $\frac{6}{16}$ novèmbris, 1643.

K. Southwell
IOURNAL
DU VOYAGE
DU CHEV.R CHARDIN
EN PERSE & AUX INDES
ORIENTALES
Par la Mer Noire & par la Colchide.
Hab: Persarum
Colch: sup: nivem ambul:
A LONDRES
Chez Moyse Pitt en Duke-street à Westmunster. 1686.

zar du Mhordar Kochon, le Garde des ſceaux de la guerre, lequel *Bazar* ſe rend au *Caravanſerai* nommé *Begum*, ou de *la Reine*, parce qu'il a été fondé par la mere de *Sephy premier*. On voit tout proche un autre *Caravanſerai*, & un *Bain*, qui portent tous deux le nom de *Payder*.

Dans les *Ecuries Royales* dont je viens de parler, il y avoit la premiere fois que j'arrivai à *Iſpahan* un *Rhinoceros*, que j'allai voir pluſieurs fois, pour en mieux prendre l'idée, & que je fis tirer par mon Peintre fort exactement à diverſes repriſes. En voici la Figure à côté : C'étoit un animal grand comme un *Bœuf* de grandeur ordinaire. Sa peau eſt d'un gris brun tirant ſur le noir, comme celle des *Elephans*, mais plus rude, & plus épaiſſe. Je n'ai point vû d'animal qui en ait une ſemblable, & cela ſe peut juger de ce qu'on ne voit point au *Rhinoceros*, comme aux autres animaux, les articulations, ni les apophyſes ou éminences des os. Cette peau eſt couverte par tout, hormis au cou & à la tête, de petits nœuds ou durillons, ſi fort ſemblables à ceux des écailles de *Tortues*, tant pour la forme, que pour la couleur, qu'à la premiere vûe on croiroit que cet animal eſt couvert d'une telle écaille ſur le corps. Cette peau fait cinq plis gros & épais, outre celui qui eſt le long du cou au deſſous des oreilles, reſſemblant à une fraiſe qui pendroit tout autour : un pli couvre toutes les Epaules juſqu'au ventre : un autre couvre le ventre & le dos entier : & trois autres couvrent les cuiſſes ; mais pliſſés en long, au lieu que les autres ſont en travers, comme on le voit dans le deſſein. La corne de cet animal, qui en eſt la partie la plus admirable, eſt preſque de la figure & de la groſſeur d'un pain de ſucre de deux livres. Sa couleur eſt de gris brun de même que la peau de la tête au-deſſus des Narines : Son muſeau eſt rond, tourné comme un bec d'Aigle, & cependant la levre au-deſſus de la bouche eſt plate & large. Il n'a que quatre dents, deux en haut & deux en bas, placées aux extremitez des machoires. Sa langue eſt courte & épaiſſe. Ses yeux ſont placés fort bas, preſque contre les levres. Sa queüe n'a pas un pied de long. Elle eſt menuë, formant huit ou dix nœuds, reſſemblant à un chapelet. Ses pieds ſont courts & épais, faits de trois fourchons, ou argots de corne ſur le devant, & de durillon ſur le derriere. On entretenoit ſi miſerablement ce pauvre animal quand je le vis, (ſon gardien ſouſtraïant ſa nourriture,) que malgré l'épaiſſeur de ſa peau, on lui voioit les côtes au-travers. J'en obſervai huit, attachées aux vertebres, qui compoſent ſon épine de dos. Les *Perſans* appellent cet animal *El kerkedon*, c'eſt-à-dire *le porte-corne*, ou *aiant corne*. La *Relation Hollandoiſe*, qui a pour titre *l'Ambaſſade de la Chine*, fait une deſcription de cet animal tout à fait fauſſe, ſur tout en ce qu'elle porte que c'eſt un des principaux ennemis de l'Elephant : car ce *Rhinoceros*-ci étoit dans une même écurie avec deux *Elephans*, & je les ai vû diverſes fois tous trois l'un près de l'autre dans la *Place Royale*, ſans ſe marquer la moindre antipathie. Un Ambaſſadeur d'*Ethiopie* avoit amené cet animal en préſent. C'eſt le païs où il y en a davantage, & je n'ai pas pû découvrir qu'il y en ait aux *Indes*. Les *Abiſſins*, ou *Abechi*, comme les *Perſans* les appellent, les apprivoiſent, & élevent au travail, comme on fait les *Elephans*. On pretend qu'aux *Indes* les Rois & Princes ſe ſervent de cornes de *Rhinoceros* à boire, à cauſe de l'antipathie qu'elle a avec le poiſon, lequel ſe reconoit en ce que la corne ſue au moindre poiſon qu'il y a dedans. Je vous aſſure que la premiere partie du conte eſt fabuleuſe, je ne ſaurcis rien dire de l'autre, n'en aïant pas vû d'épreuve.

Quand on a paſſé ce Quartier-là on entre dans celui de *Nimaourde*, qui eſt un des plus fameux & des plus peuplés d'*Iſpahan*. Ce qu'il y a de remarquable eſt la *Ruë choumalou* : La *Moſquée* de *Zoulfogar*, qui eſt le nom du *Sabre d'Aly* : un *Bain* & un *Hôtel*, qui porte le nom de *Kaſſé-trache*, c'eſt-à-dire *le Barbier du corps*, qui eſt celui qui fait le poil au Roi, ce qui eſt un office conſiderable : le Logis de *Cheib Mirza* Vizir du Païs de *Karaoloss* : la *Ruë Neuve*, où eſt une maniere de *Convent* pour les *Dervîches* de la ſecte des *Souphis*. On l'appelle *le Repoſoir des Derviches Soufis* : le *Bain Lavandié* : la *Ruë des Juifs*, où on montre une de leurs *Synagogues* : le *Bazar d'Aramené*, & *le Caravanſerai d'Abas* : c'eſt le Prince premier du nom, qui le fit conſtruire, & c'eſt un des beaux *Caravanſerais* de la ville. On fait obſerver à l'entrée la *Pierre* ſur laquelle ce Monarque fit mettre en piéces un fameux ſcelerat qui enlevoit les garçons pour les proſtituer : Il ſe tenoit-là le long du jour, & quand il en appercevoit quelqu'un qui lui plaiſoit il l'enlevoit adroitement ; & l'aiant gardé toute la nuit, il le remenoit au point du jour en quelque endroit écarté, afin qu'on ne pût ſavoir où il avoit été. *Abas le Grand* aïant apris la choſe, & que les avis & les menaces du voiſinage n'y avoient

Unfortunately, he limited his research only to this part of the Ottoman Empire and did not explore any of the countries of the Middle East.

Considerably more famous than Wheeler's work is Sir John Chardin's book on the East. Chardin, a French Protestant, travelled to Persia in search of jewellery. After two journeys to the East, where he lived for over a decade, he returned to Europe and settled in England to escape persecution in France. He was knighted by Charles II in 1681, and the following year he was elected a fellow of the Royal Society. In 1685 he published the first part of his *Travels in Persia* (*Voyages du Chevalier Chardin en Perse, et autres lieux de l'Orient*), but did not complete his work till 1711. The book, which was praised by Voltaire and Gibbon, offers a detailed account of contemporary Persia and has chapters on trees, plants, drugs, flowers, fish, birds and mammals. Chardin's descriptions of natural history complete his picture of the East, but are of limited scientific significance.

The case of Jean de Thévenot, a French traveller and botanist who travelled extensively in Syria, Palestine, Egypt, Mesopotamia and Persia during the 1650s and 1660s, is more difficult to assess. Although reputedly a competent naturalist who put together an important collection of plants during his travels, he made no contribution to the natural history of the Near or Middle East. He was an avid collector but unfortunately only the Indian volume of his five herbariums ever reached France. Indeed, the almost complete absence of botanical descriptions in his accounts of his journeys is striking. In the occasional but superficial descriptions of the landscape and of the vegetation of the East, Thévenot merely mentions the existence of gardens, trees, reeds, rocks, rivers and woods, but very rarely reports on his botanical observations. In the first part of his travelogue *Relation d'un voyage fait au Levant* (1664) he refers the reader to Alpini's work for the plants of Egypt,[8] and of the over 400 pages of the *Suite du voyage de Levant* (1674), describing his journey to Persia, only five pages are devoted to plants and their fruits.[9] Although Thévenot had drawn up lists of all the plants he had collected, many of them genuine new discoveries, these were only published in 1906, and by then they did little to increase the botanical knowledge of the Middle East. Thévenot's collecting of plants might well suggest that he contemplated writing a work on the flora of the East once he had returned to France. This was not to be. He died in Armenia in 1667 on his way back to Europe.

Unlike the travellers of the seventeenth century, the eighteenth-century naturalists who visited the Near East contributed significantly to the European knowledge of its natural history. At the turn of the century one of Europe's greatest botanists undertook a journey to the Levant. Joseph Pitton de Tournefort, who had studied botany in Montpellier, had made a name for himself with his early work on the French flora, and had consequently been appointed professor of botany at the Jardin des Plantes in Paris in 1683. He there worked on a new method of classification of plants based on the structure of their flowers and fruits, and he distinguished between generic and specific rank. Linnaeus was to adopt many of Tournefort's genus names in his binominal system. In 1700 Tournefort received orders from King

8. '[J]e ne parle point de plusieurs autres plantes [...] qui sont décrites dans Prosper Alpinus', Jean de Thévenot, *Relation d'un voyage fait au Levant*, Paris 1664, p. 494.
9. Jean de Thévenot, *Suite du voyage de Levant*, Paris 1674, pp. 224–30.

Louis XIV to explore the natural history and geography of the Levant, and gather information about its commercial possibilities. For two years he travelled with the draughtsman Claude Aubriet and the German botanist Andreas Gundesheimer from Istanbul to Tiflis in Georgia, and crossed Turkey back to Izmir, collecting over 3,000 plants on the way, of which over a third were new species. The account of these travels was published posthumously in 1717 as *Relation d'un voyage au Levant*. It included detailed drawings of new plants as well as a series of fine coloured plates.

The first half of the eighteenth century saw three further important contributions to the study of natural history by Thomas Shaw, Richard Pococke and Fredrik Hasselquist.

Shaw was a clergyman who travelled to Algiers in 1720 to take up a post as chaplain to the factory of English merchants. There being few calls for the duties of his profession, he dedicated most of his time to touring and studying the north of Africa, travelling as far as Egypt and Syria. Shaw had had a classical education at The Queen's College, Oxford, and was particularly interested in Greek and Roman antiquity in the Arab world. Moreover, he also studied the natural history of the countries he visited, and his *Travels, or Observations Relating to Several Parts of Barbary and the Levant* (1738) contains lists of over 400 plants, corals, fish and shells. Shaw was a good observer and his book was illustrated with pictures drawn by himself. The part on Egypt contains interesting observations on the plants and animals of that country. He notes that the crocodile 'so rarely appears below the cataracts, that the sight of it is as great a curiosity to them [the Egyptians] as to the Europeans'.[10] Shaw also included a detailed description and a drawing of the skeleton of an embalmed ibis, which he had brought back from Egypt. However, his acceptance of a report of how an Italian kept a couple of Egyptian vipers for 'five years in a large crystal vessel, without any visible food'[11] shows some of the weaknesses of Shaw's work.

Although Shaw's work was criticized by Richard Pococke, the *Travels* enjoyed lasting popularity. A third edition appeared as late as 1808, more than half a century after the author's death, and the book was translated into Dutch, German and French. After the publication of his *Travels* Shaw was elected Principal of St Edmund Hall at the University of Oxford and was appointed Regius Professor of Greek.

Pococke is famous for his voluminous *Description of the East*, based on his travels in Palestine, Syria, Egypt and Mesopotamia in 1737–8. The book contains chapters with detailed information about the plants and animals he encountered. Unfortunately, his descriptions of animals are uneven in quality, and he often compares his observations with those of Herodotus. Of the animals in Egypt, for example, he mentions in one breath hyenas, hares, foxes, camels and horses, and he distinguishes only two species of snakes, four species of lizards, and a handful of birds.[12] However, none of these animals is described scientifically. But if Pococke was a poor zoologist, his botanical knowledge was far from superficial and much superior to that of the average traveller to the East. The descriptions of the countries he visited are interspersed with botanical observations, mixing technical description

NEXT OPENING
Opening of third book of Jean de Thévenot's *Relation d'un voyage fait au Levant* with woodblock print of 'Kherzehreh' (*Calotropis procera*).
Jean de Thévenot, *Relation d'un voyage fait au Levant* [*with*] *Voyages … contenant la relation de l'Indostan, des nouveaux mongols, et autres peuples et pays des Indes*, Paris 1664, pp. 230–1
The Arcadian Library, 08944

FOLLOWING OPENING
Woodblock print of Tournefort's Eastern Poppy ('Papaver Orientale hirsutissimum, flore magno').
Joseph Pitton de Tournefort, *Relation d'un voyage du Levant*, Paris 1717, vol. 3, p. 127
The Arcadian Library, 13192

10. Thomas Shaw, *Travels, or Observations Relating to Several Parts of Barbary and the Levant*, London 1738, p. 427.
11. Shaw, *Travels*, p. 429.
12. Richard Pococke, *A Description of the East and Some Other Countries*, 2 vols., London 1743–5, IV.ix.

Le Cherzehre.
Le Connar.

LIVRE TROISIE'ME.

DV PAYS DE SCHIRAS ET AVTRES lieux qui sont sous la domination du Roy de Perse

CHAPITRE PREMIER.

De la route d'Ispahan à Schiras.

APRES vn sejour de prés de cinq mois à Ispahan, je me disposay à passer outre: Et j'en partis le vingt-quatriesme Février de l'année mil six cent soixante & cinq, avec vne caravane, dans laquelle il y avoit environ cinquante mules, dont vne bonne partie appartenoit à Monsieur Tavernier, & le reste à des Armeniens, qui se servoient de l'occasion de nostre départ. Nous prîmes des mules pour nos hardes, à raison de cinq abassis, pour cent mans de Tauris, dont deux ne font qu'vn man d'Ispahan, & vne mule porte cinquante à soixante mans de Tauris. Pour nos personnes nous avions nos chevaux; aussi bien les muletiers faisoient difficulté de loüer des mules de montures: ils furent pourtant obligez d'en donner vne à mon valet, qui mît dessus vne partie de mes hardes; car ils ne comptent vne personne que pour trente mans, y comprenant quatre ou cinq mans de hardes. Nous partîmes donc de Giolfa vn Mardy à midy; nous passâmes devant Hezar Dgirib, & nous allâmes droit au levant. Nous campâmes à vne heure aprés midy contre vn Kervanseray appellé Tahht Poulad & Babaruk, qui est proche le cimetiere des Mahometans.

Départ d'Ispahan.

Tahht Poulad: Babaruk.

dans les flâmes avec leurs femmes & leurs enfans. Abraham y trouva beaucoup d'or & des ferrements que le feu n'avoit pû devorer. Il en fit sortir plusieurs chevaux & autres bêtes de somme. Zonare raconte à peu près la même chose de la destruction d'Artze, mais il ne parle pas de Theodosiopolis. Cet auteur assûre seulement qu'Artze etoit sans muraille, & que ses habitans en avoient fortifié les avenuës avec du bois, je crois qu'ils consumérent tout celui qui étoit aux environs, car depuis ce temps-là l'espece s'en est perduë. Comme la Place fut réduite en cendres, & que ce passage est absolument nécessaire pour le commerce, il y a beaucoup d'apparence que les restes de ces pauvres habitans, & les marchands étrangers qui s'y vinrent établir dans la suite, pour ne pas tomber dans un pareil malheur, se retirérent à Theodosiopolis qui en étoit tout près, suivant Cedren.

Les Turcs à qui peut-être le nom de *Theodosiopolis* parut trop long & trop embarrassant, donnérent le nom d'*Artzé-rum* à cette Place, c'est à dire *Artze des* [a] *Grecs* ou *des Chrêtiens*, car *Rum* ou *Rumili* signifie en langue Turque *la Romanie* ou *la terre des Grecs*. Ils distinguent *la Romelie* ou *Rumili* en celle d'Europe & en celle d'Asie, ainsi d'*Artzé-rum* on a fait *Arzerum*, & *Erzeron*, comme prononcent la plûpart des Francs. Il ne faut pas confondre cette ville de *Theodosiopolis* avec une autre ville de même nom, qui étoit sur le fleuve *Abhorras* en Mesopotamie, & que l'Empereur Anastase avoit fait revêtir de fortes murailles, comme l'assûre Procope. Ce même auteur fait mention de la *Theodosiopolis*

[a] Ρώμαιοι.

Papaver Orientale hirsutissimum, flore
magno Coroll Inst. Rei herb. 17.

nosa, alopecuroides, flore rubello, Sicula Bocc. App. ad Mus.

39 Anonis viscosa, spinis carens, lutea, latifolia Bot. Monsp. Anonis non spinosa, capreolis donata C. B. P. 389.

40 Antirrhinum latifolium, flore rubro, rictu luteo Boerh. Ind. Alt. 233.

41 Aphaca Lob. Ic. 70.

42 Apium procumbens, crassiore folio.

43 Apocynum erectum, incanum, latifolium, Malabaricum, floribus ex albo suave-purpurascentibus Par. Bat. 28. Boerh. Ind. Alt. 313. *Copiose crescit in vallibus prope montem Sinai.*

44 Apocynum frutescens, folio subrotundo, minore, siliquis strictissimis.

45 Arbor Judæ Dod. Pempt. 786.

46 Arbutus, Comarus Theophrasti J.B. I. 83.

47 Aristolochia rotunda J.B. III. 559. *Borustum* Arabum.

48 Aristolochia clematitis recta C. B. P. 307.

49 Aristolochia Cretica, scandens, altissima, Pistolochiæ foliis Cor. 8. Aristolochia clematitis serpens C. B. P. 307.

50 Arum humile, Arisarum dictum, latifolium, pistillo brevi, hirsuto, incurvo H. L. Arisarum latifolium majus C. B. P. 196.

51 Asparagus sive Corruda, spinis biuncialibus, binis.

52 Asplenium sive Ceterach J. B. III. 749.

53 Aster conyzoides, foliis angustis, crenatis.

54 Aster maritimus, flavus, folio in summo obtuso H. L. Flor. I. 23. Boerh. Ind. Alt. 95. n. 15.

55 Aster pratensis, autumnalis, Conyzæ folio I. R. H. 482. Helenium pratense autumnale, Conyzæ foliis, caulem amplectentibus Comm. Ac. R. Sc. Ann. 1720. p. 303. n. 11.

56 Asteriscus perennis, foliis longis, angustis.

57 Asteriscus annuus, foliis ad florem rigidis I. R. H. 497. Aster Atticus Massilioticus Tab. Ic. 861.

58 Asteriscus annuus trianthophorus, *Craffas* Arabibus dictus. *Folia Chamæmeli. Calyx e squamis tenuibus, albo virentibus, constat. Semiflosculi sinuati sunt: Crenas laterales longiores, mediam breviorem habet. Suaviter olet.*

59 Asteriscus maritimus, perennis, patulus I. R. H. 498. Aster supinus, lignosus, Siculus, Conyzæ odore Bocc. Mus. P. II. 161.

60 Astragaloides Lusitanica I. R. H. 399. Astragalus Bœticus Clus. H. CCXXXIII. *Foole el Haloufe* (s. Faba Apri) Arabum.

61 Astragalus Africanus luteus odoratus Bot. Monsp. Astragalus perennis foliis hirsutis, caule recto aphyllo, flore ochroleuco, odoratissimo H. Ox. II. 203. *Caroube el Maizah* (s. Siliqua Caprarum) Arabum.

62 Astragalus annuus, angustifolius, flosculis subcæruleis, cauliculis adhærentibus I. R. H. 416. Astragalus siliquis & foliis hirsutis, floribus parvis H. Ox. II. 119.

63 Astragalus Bœticus sive Securidica Sicula, siliquis foliaceis Bocc. Rar. p. 7. Tab. 4. Astragalus triangularis Munt. Phyt. Tab. 10.

64 Astragalus luteus, annuus, Monspeliacus, procumbens H. Ox. II. 108. Securidaca lutea minor, corniculis recurvis C. B. P. 349.

65 Astragalus Monspessulanus J. B. II. 338. I. R. H. 416.

66 Astragalus pumilus, siliqua Epiglottidis forma I. R. H. 416.

67 Astragalus supinus, siliquis villosis, glomeratis I. R. H. 416.

68 Astragalus tenuifolius, flore sulphureo, siliquis tenuiter recurvis.

69 Atractylis multiflora cærulea Comm. Ac. R. Sc. An. 1718. p. 171. n. 8. Carthamus aculeatus Carlinæ folio, flore multiplici, veluti umbellato Cor. 33.

70 Atriplex angusto, oblongo, folio C. B. P. 11[illegible] H. Ox. II. Tab. 32. Sect. 5.

71 Atriplex maritima, Hispanica, frutescens & procumbens I. R. H. 505. Hort. Elth. 46. Fig. 46.

72 Atriplex maritima pumila, Arabica, foli[illegible] villosis, subrotundis. *Folia unguis equini figura.*

73 Atriplex olida, maritima, pumila, procumbens.

74 Azedarach Dod. Pempt. 848. I. R. H. 61[illegible] *Eleah* Arabum.

75 Balsamita Chrysanthemi segetum folio[illegible] disco amplo.

76 Balsamita foliis Agerati Comm. Ac. R. S[illegible] Ann. 1719. p. 280. n. 2. Bellis spinosa, foliis Agerati C. B. P. 260. Bellis spinosa Pr. Alp. Ex. 32[illegible]

77 Blattaria flore albo J. B. III. 874.

78 Blattaria magno flore C. B. P. 241.

79 Bryonia aspera sive alba, baccis rub[illegible] C. B. P. 297.

80 Buglossum angustifolium Lob. Ic. 76.

81 Buglossum Creticum majus, flore cærule[illegible] purpurascente H. R. Par. I. R. H. 134.

82 Buglossum Lusitanicum, Echij folio, und[illegible] lato I. R. H. 134.

83 Buglossum luteum, annuum, minimum I. R. H. 134. Anchusa lutea minor J. B. III. 583.

84 Buglossum radice rubra, sive Anchusa vu[illegible]gatior, floribus cæruleis I. R. H. 134. Anchusa p[illegible]niceis floribus C. B. P. 255.

85 Buglossum sylvestre majus nigrum C. B. 256. Borrago sylvestris annua di Candia Zan. H.[illegible]

86 Bulbocastanum minus C. B. P. 162.

87 Bulbocastanum tenuiter inciso folio Lusi[illegible]nium Vir. Lus. I. R. H. 307.

88 Bulbocodium crocifolium, flore parv[illegible] violaceo I. R. H. Cor. 50. Sysirynchium The[illegible]phrasti Col. Ec. I. 328.

89 Bupleurum perfoliatum, rotundifolium, [illegible]nuum I. R. H. 310. Perfoliata vulgatissima [illegible] arvensis C. B. P. 277.

90 Bupleurum arborescens Salicis folio I. R. 310. Seseli Æthiopicum fruticosum folio Pe[illegible]clymeni J. B. III. p. 2. 197.

91 Bursa Pastoris hirsuta, Erucæ flore, [illegible] prominente. *Folia oblonga, serrata, caulem [illegible] plectentia. Siliquæ hirsutæ, interdum ex adv[illegible] positæ, brevibus pediculis in spicam digestæ, Bu[illegible] Pastoris figura, sed majores & altius sinuata. Sep[illegible] medium Geranij seminis instar exporrectum.*

92 Cakile maritima, angustiore folio Cor.

93 Cakile maritima, ampliore folio Cor. Eruca maritima, latifolia, Italica, siliqua ha[illegible] cuspidi simili C. B. P. 99.

94 Calamintha Cretica, angusto, oblon[illegible] folio I. R. H. 194.

95 Calcitrapa flore sulphureo, procumbe[illegible] caule non alato. Jacea Cichorii folio, flore lut[illegible] capite spinoso Bocc. Rar. 15. Jacea orient[illegible] spinosa, folio Erysimi, flore luteo Boerh. [illegible] Alt. 141. *In junioribus capitulis, spinæ super[illegible] reliquis longiores sunt, & castanei coloris.*

96 Calcitrapa laciniata, multiflora, min[illegible] flore, albicante Comm. Ac. R. Sc. Ann. 1[illegible] n. 165. Carduus orientalis Calcitrapæ folio, [illegible] minimo Cor. 31. Jacea minor &c. Pluk. A[illegible] 192. Tab. 39. f. 4.

97 Calcitrapa lutea alato caule, capite e[illegible]phoro Comm. Ac. R. Sc. Ann. 1718. p. 166. n[illegible]

98 Calcitrapa vulgaris, lutea, alato caule Co[illegible] Ac. R. Sc. Ann. 1718. p. 166. n. 21. Car[illegible] stellatus luteus, foliis Cyani C. B. P. 387. I[illegible] H. 440.

99 Calcitrapoides Rapi folio, alato c[illegible] flore purpureo coronato Comm. Ac. R. Sc. [illegible] 1718. p. 168. n. 1.

100 Calcitrapoides Sonchi folio, capite m[illegible] turbinato Comm. Ac. R. Sc. Ann. 1718. p. n. 10. Carduus stellatus, latifolius, caule[illegible] C. B. P. 387.

101 Calcitrap[illegible]

Exc. p. 38.

5 Acetosa &c.
58. Asteriscus &c.
72 Atriplex &c.
176. Conyza &c.
146 Chondrilla &c.
Auspiciis Rev. et Honoratis Dom. Domini
JACOBI BEAUCLERC Canonici Windesoriensis.

PREVIOUS OPENING
Thomas Shaw, plate with plants from Egypt.
Travels, or Observations Relating to Several Parts of Barbary and the Levant, Oxford 1738, facing p. 38
The Arcadian Library, 08358

Georg Dionysius Ehret, engraving of *Palma thebaica*.
Richard Pococke, *A Description of the East and Some Other Countries*, London 1743–5, plate LXXII, II.280
The Arcadian Library, 09032

with information on how natural products were used by the local populations. For example, he drew attention to the fact that the 'Palma Thebaica' or gingerbread tree had never been adequately described or depicted.[13] During all his travels he collected specimens of plants. He either pressed them,[14] or sent seeds to Philip Miller at the Chelsea Physic Garden in London, who grew plants from them. It was Miller too who provided Pococke with long lists of the collected species, which Pococke again integrated with detailed scientific descriptions, based on his personal observations. The botanical illustrator Georg Dionysius Ehret, who had just completed his work with Linnaeus on the *Hortus Cliffortiana*, a publication which has been seen as 'botanical art of a wholly new order of elegance and accuracy',[15] was engaged to make drawings of those species that had not yet been depicted. The preface to the French edition of the *Description* (1772) drew special attention to his botanical achievements: although Pococke was no doubt an erudite observer, his botanical qualities required special praise. Botany is not a 'sedentary and lazy science' which one can master in the cabinet as one can history, it was argued, but a science which requires 'crossing mountains, climbing steep rocks and venturing to the edges of precipices'. 'Our English traveller', it concluded, 'must have been indefatigable for having survived the fatigues and dangers of the pursuit of plants.'[16]

A decade after Pococke the 27-year-old Swede Fredrik Hasselquist set out for the Middle East. Always eager to send young students around the world to collect plants and animals, Linnaeus once stated during a lecture that Palestine was among those countries in the world about whose natural history we were still relatively ignorant. The remark was not lost on Hasselquist, who had enrolled at the University of Uppsala to study Medicine and Natural History. He made up his mind to explore the Holy Land, and to prepare himself for such a journey he took lessons in Arabic and found the necessary funding. He sailed to Izmir in 1749, and to Alexandria and Cairo in 1750. In his lively diary (published as *Iter Palaestinum eller Resa til Heliga Landet* in 1757) he intersperses botanical and zoological information with accounts of local customs, and adds how he obtained specimens of plants and animals. Soon after arriving in Egypt, for example, he was struck by snake charmers. At the beginning of July he arranged to have snakes brought to him, as 'the great heats bring [...] forth these vermin', and have them put in spirits. A woman who arrived with specimens of 'the common Viper, the Cerastes of Alpin, Jaculus, and an Anguis marinus', filled the Europeans with 'consternation' as

she handled the most poisonous and dreadful creatures alive and brisk, without their doing or even offering to do her the least harm. When she put them into the bottle where they were to be preserved, she took them with her bare hands, and handled them as our ladies do their laces. She had no diffculty with any but the Viperae Officinales, which were not fond of their lodging. They found means to creep out before the bottle could be corked. They crept

13. Pococke, *A Description of the East*, IV.iii, V.xvii.
14. One volume of pressed flowers is preserved in the Heritage Library in Qatar. See Alastair Hamilton, *The European Legacy: Arab and Islamic Culture in the Heritage Library of Qatar Foundation*, Doha 2006, pp. 50–1.
15. Mark Griffiths, 'Clifford's Banana: How Natural History was Made in a Garden', in *The Linnaean Collections*, ed. Mary J. Morris and Leonie Berwick, London 2007, p. 30.
16. Pococke, *Voyage de Richard Pockocke en Orient, dans l'Égypte, l'Arabie, la Palestine, la Syrie, la Grèce, la Trace, &c. &c. &c.*, 7 vols., Paris 1772, I.vi.

PALMA THEBAICA

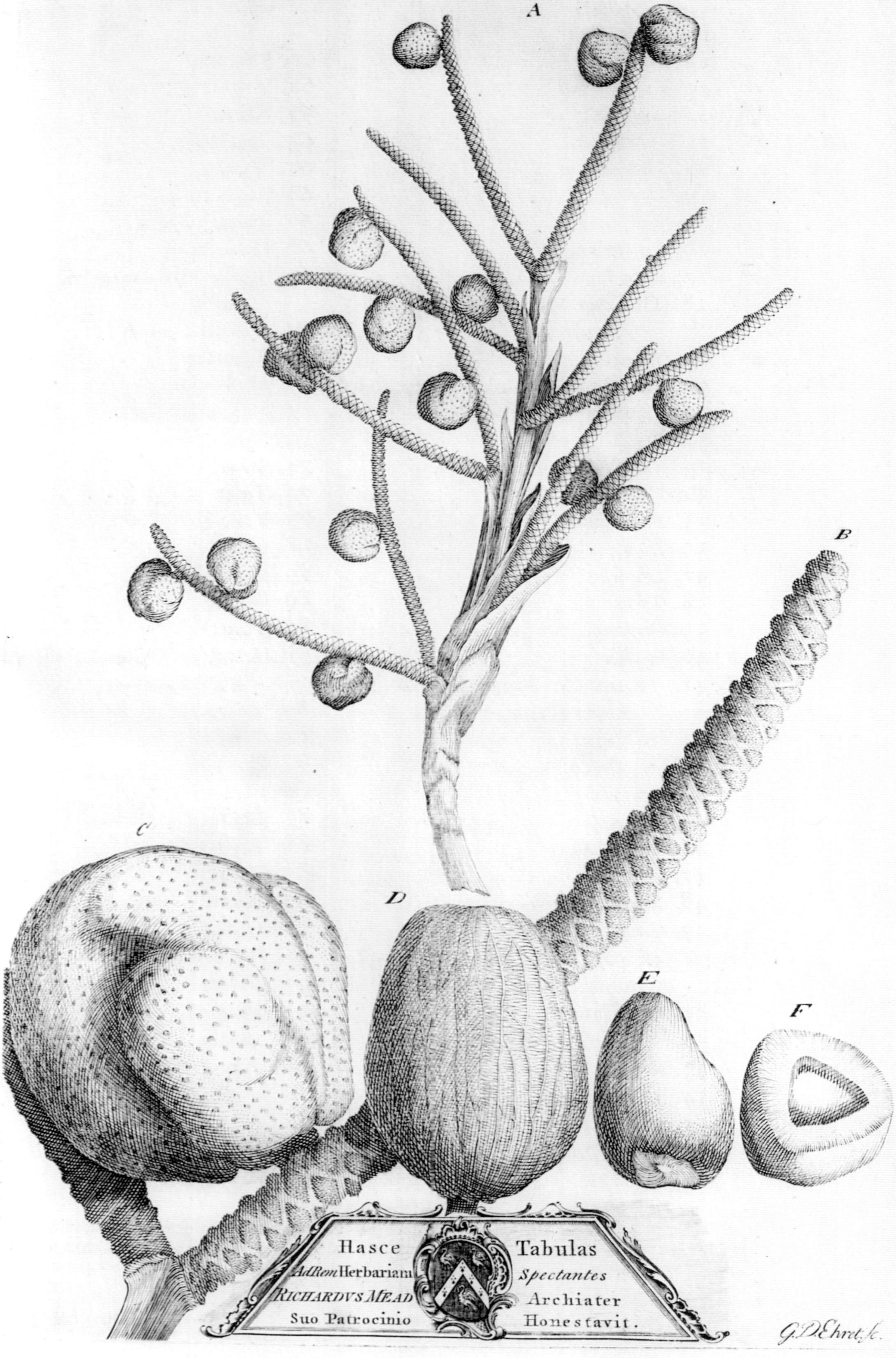

G. D. Ehret. sc.

Palma Thebaica

over the hands and bare arms of the woman, without occasioning the least fear in her: she with great calmness took the snakes from her body, and put them into the place destined for their grave.[17]

Georg Dionysius Ehret, engraving of *Palma thebaica*. Richard Pococke, *A Description of the East and Some Other Countries*, London 1743–5, plate LXXIII, II.281 The Arcadian Library, 09032

Hasselquist wondered how the woman could handle snakes like this, and asked the Arab onlookers, considering it 'worthy the endeavours of all naturalists' to shed some light on this practice. The Arabs told him of some 'unknown art' possessed only by certain families. Before going out to catch serpents, the snake charmer ate a dish of them raw or boiled in a broth and received a blessing in a ceremony from their priest who 'spits on them several times with certain gestures'. Hasselquist, however, suspected that they anointed themselves with some kind of herb, and from the explanation offered him by the Arabs he deduced that 'We see by this, that they know to make use of the same means used by other nations; namely, to hide under the cloak of religion, what may be easily and naturally explained,'[18] a sceptical and realistic stand he shared with his younger colleague Pehr Forsskål, another of Linnaeus' disciples.

From Egypt Hasselquist sent numerous botanical and zoological observations to the Royal Academies of Uppsala and Stockholm, which were duly published in their transactions. In 1751 he left Egypt for Jaffa and Jerusalem, and reached Izmir by the end of the year. Having long suffered from weak lungs, he died on 9 February 1752, just before he could sail back to Sweden. When his extensive collections of pressed plants, fish, snakes, and insects reached Sweden, Linnaeus was 'astonished at the sight of so many unheard of curiosities'.[19] Hasselquist's diaries were published posthumously by Linnaeus, who also edited the scientific results of the expedition. It was he who placed Hasselquist's species in their proper genera, giving them correct names according to the principles of the nomenclature proposed in his *Systema naturae*, describing 52 plants and 137 mammals, birds, reptiles, fish, insects and worms.

Modern botanical and zoological exploration of the Levant in terms of Linnaean nomenclature started with an ambitious expedition to Arabia Felix (generally located in Southern Arabia) sponsored by the Danish Crown. This first full-scale scientific expedition consisted of five members—the German mathematician Carsten Niebuhr, the Danish philologist Frederick Christian von Haven, the Swedish naturalist Pehr Forsskål, the Danish physician Christian Carl Kramer, and the German draughtsman Georg Wilhem Baurenfeind. They reached Cairo in 1761, and stayed there for a year and half, exploring the area. At the beginning of 1763 they sailed down the Red Sea, and set foot on Yemenite soil. The scientific results of these travels were published a decade later by Niebuhr in his *Beschreibung von Arabien* (1772), while a personal account of the journey was given in his three-volume *Reisebeschreibung nach Arabien und andern umliegenden Ländern* (1774, 1778,

17. Fredrick Hasselquist, *Voyages and Travels in the Levant; In the Years 1749, 50, 51, 52. Containing Observations in Natural History, Physick, Agriculture, and Commerce: Particularly on the Holy Land, and the Natural History of the Scriptures*, London 1766, p. 63.
18. Hasselquist, *Voyages and Travels in the Levant*, p. 65.
19. Hasselquist, *Voyages and Travels in the Levant*, p. vii.

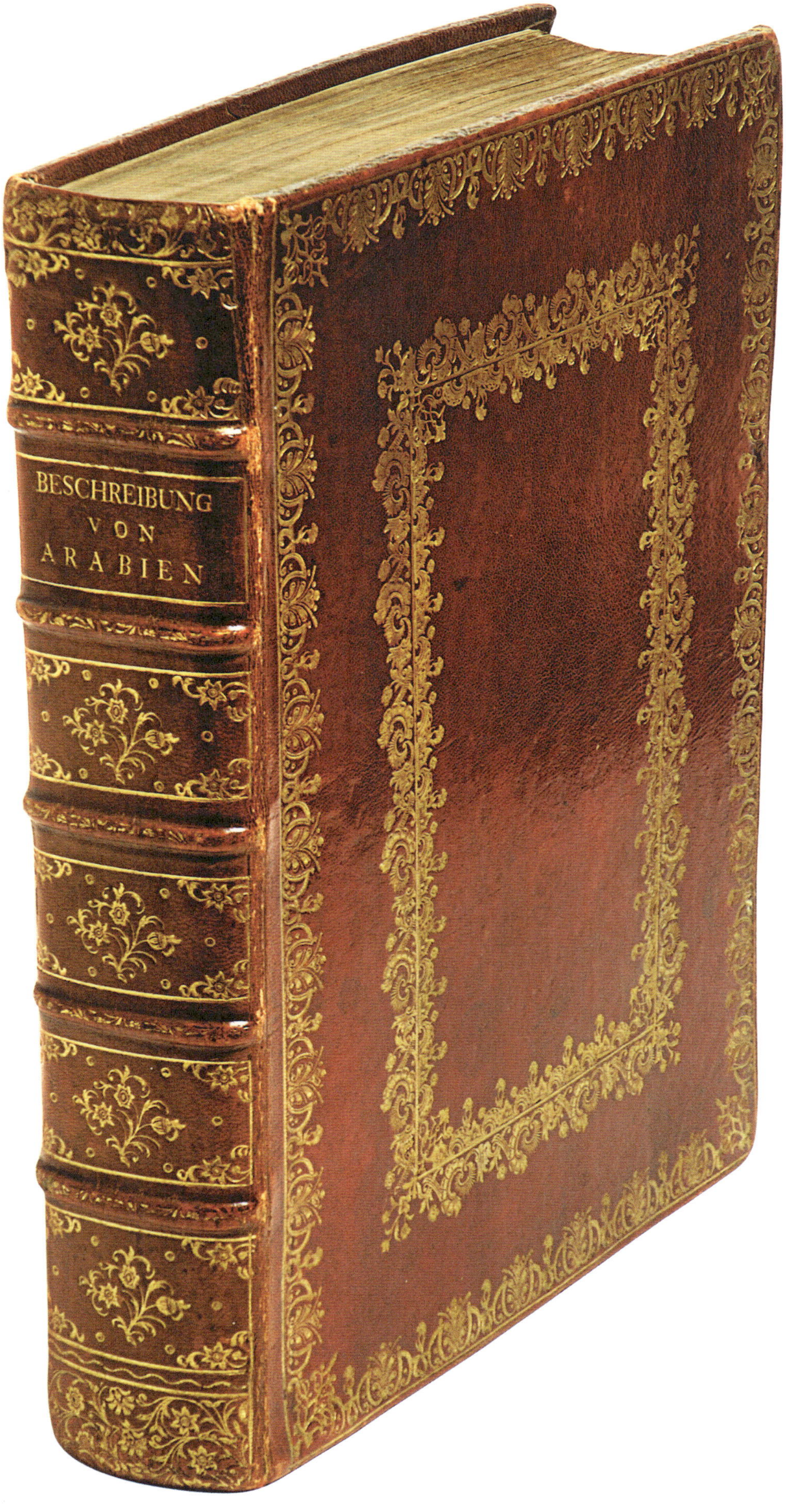
BESCHREIBUNG
VON
ARABIEN

1837). In 1775 he also edited Forsskål's *Flora Aegyptiaca-Arabica* and *Descriptiones animalium*, the first Linnaean works on the natural history of Egypt, the Red Sea and the Yemen. Forsskål's contribution to European knowledge of the flora and fauna of the countries he visited was of enduring importance, and many of the new specimens he collected and described are the type specimens of plants and animals occurring throughout Arabia and Africa.

Carsten Niebuhr, *Beschreibung von Arabien: Aus eigenen Beobachtungen und im Lande selbst gesammleten Nachrichten abgefasset*, Copenhagen 1772 The Arcadian Library, 10525

An expedition of even greater dimensions followed twenty years later when Napoleon Bonaparte took over a hundred scientists and artists with him on his Egyptian campaign. Although the military side of the campaign was ultimately a failure, the scientific side was not. The naturalists carried out work on the Egyptian flora, fauna and geology. They included the renowned mineralogists Déodat de Dolomieu and Pierre Louis Antoine Cordier, the chemist Claude-Louis Berthollet, the zoologists Étienne Geoffroy Saint-Hilaire and Jules-César Savigny and the botanist Alire Raffeneau-Delile. If the first three were already scientists of repute, the last three all made their names through their work in Egypt. For three years they studied the natural history of the country, exploring the Nile Delta, the Red Sea, the desert and the Upper Nile. Savigny contributed significantly to the knowledge of the invertebrates of the Red Sea, Geoffroy did important work on the fish of the Nile, and Delile put together a fine collection of the plants of Lower and Upper Egypt. Their scientific results were published in the natural history volumes of the sumptuously illustrated *Description de l'Égypte*.

The greatest number of specimens was collected by the Prussian expedition of the 1820s. The botanist Christian Gottfried Ehrenberg and the zoologist Wilhelm Hemprich explored for six years extensive parts of Egypt, Lebanon, Libya, Sinai, the Red Sea, Saudi Arabia and the Yemen, getting as far as Dongola and Eritrea. Notwithstanding countless difficulties, Ehrenberg and Hemprich collected tens of thousands of plants, and well over 30,000 specimens of animals were sent back to Berlin. Many of the new species were described in *Symbolae physicae*, a work with fine colour plates which was to rival the French *Description*.

Apart from the obvious scientific, military, topographical and commercial interests involved, more subtle imperialistic designs can be discerned in the histories of the great scientific expeditions to the Arab world. Edward Said perceived an influential textual tradition in European attitudes to the Orient, 'as something one read about and knew through the writings of recent as well as classical European authorities'.[20] In *Orientalism* (1978) he posited that the modern keynote of the relationship between the Near East and Europe was set by the Napoleonic invasion of Egypt, an 'invasion which was in many ways the very model of a truly scientific appropriation of one culture by another', and that this was nowhere symbolized so well as in its 'collective monument of erudition', the *Description de l'Égypte*.[21] Sixteen years later he returned to the importance of the *Description* as signalling the beginning of a modern European imperial appropriation of the Orient in *Culture and Imperialism*,[22]

20. Said, Edward W., *Orientalism*, Harmondsworth 2003, p. 80.
21. Said, *Orientalism*, p. 42.
22. Said, Edward W., *Culture and Imperialism*, London 1994, pp. 37–40.

Beschreibung
von
Arabien
Aus eigenen Beobachtungen
und im Lande selbst gesammleten Nachrichten
abgefasset
von
Carsten Niebuhr.

Kopenhagen, 1772.
Gedruckt in der Hofbuchdruckerey bey Nicolaus Möller.
In Leipzig zu bekommen bey B. C. Breitkopf und Sohn.

Title page of Carsten Niebuhr, *Beschreibung von Arabien: Aus eigenen Beobachtungen und im Lande selbst gesammleten Nachrichten abgefasset*, Copenhagen 1772
The Arcadian Library, 10525

Carsten Niebuhrs

Reisebeschreibung

nach

Arabien und andern umliegenden Ländern.

Erster Band.

Kopenhagen,

Gedruckt in der Hofbuchdruckerey bey Nicolaus Möller,

1774.

Title page of volume I of Carsten Niebuhr, *Reisebeschreibung nach Arabien und andern umliegenden Ländern*, Copenhagen 1774
The Arcadian Library, 12540

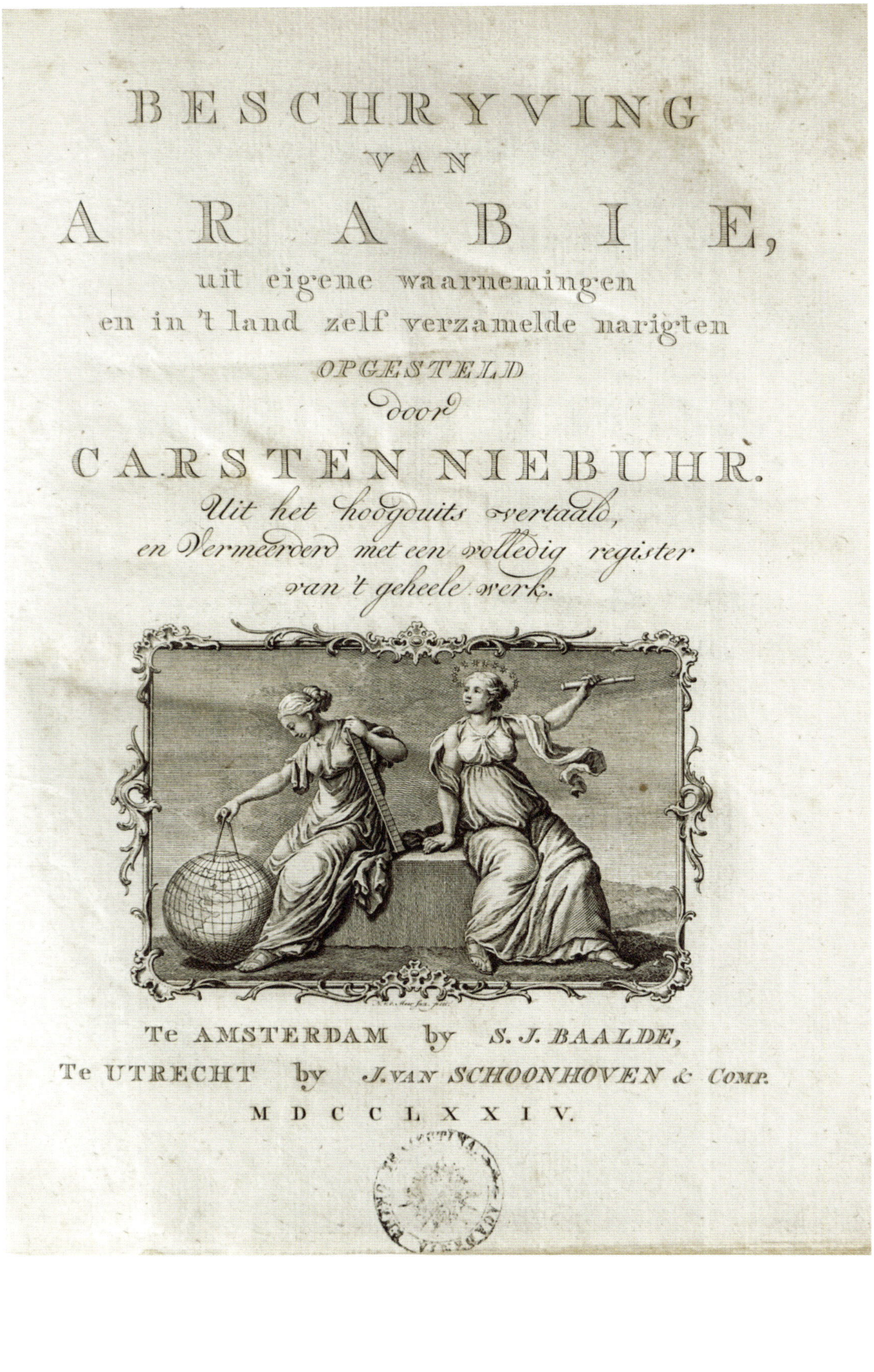

BESCHRYVING
VAN
ARABIE,
uit eigene waarnemingen
en in 't land zelf verzamelde narigten
OPGESTELD
door
CARSTEN NIEBUHR.
Uit het hoogduits vertaald,
en Vermeerderd met een volledig register
van 't geheele werk.

Te AMSTERDAM by S. J. BAALDE,
Te UTRECHT by J. VAN SCHOONHOVEN & COMP.
MDCCLXXIV.

Title page of the Dutch translation of Carsten Niebuhr's *Beschreibung von Arabien*.
Carsten Niebuhr, *Beschryving van Arabie, uit eigene waarnemingen en in 't land zelf verzamelde narigten*, Amsterdam 1774
The Arcadian Library, 08267

VOYAGE
EN
ARABIE
& en d'autres Pays circonvoisins,
par
C. NIEBUHR.
TOME PREMIER.
Traduit de l'Allemand.

À AMSTERDAM chez *S. J. BAALDE,*
À UTRECHT chez *J. VAN SCHOONHOVEN & COMP.*
M D C C L X X V I.

Title page of the French translation of Carsten Niebuhr's *Reisebeschreibung*.
Carsten Niebuhr, *Voyage en Arabie et en d'autres pays circonvoisins*, Amsterdam and Utrecht 1776
The Arcadian Library, 12541

while in his afterword to the 1995 edition of *Orientalism* he once more underscored that 'the innocent scholarly endeavour' of the *Description* 'can never unilaterally be detached from the general imperial context that begins its modern global phase with Napoleon's invasion of Egypt in 1798'.[23]

Said's imperial analysis extends to the French naturalists' endeavours in Egypt as well. He writes, for example, that in reading the *Description* it is possible 'to feel oneself as a European in command, almost at will, of Oriental history, time, and geography; [...] to divide, deploy, schematize, tabulate, index, and record everything in sight [...] and, above all, to transmute living reality into the stuff of texts'.[24] Although Said's views are helpful in reading the scientific results of the French expedition to Egypt from a Western imperial perspective, I think he is wrong about seeing the *Description* in this sense as the manifesto of modern Orientalism. I would argue, rather, that the scientific appropriation of the Orient (as indeed of any other part of the world outside Europe) started with Linnaean taxonomy, and that it was Forsskål's work in Egypt, Arabia and the Yemen that first subjected the natural history of the Near East to a modern Linnaean system.[25]

Linnaeus' naming project was ideologically and culturally charged. His attempt to index all the species of the world in a single comprehensive nomenclatural system was meant to reflect the perfection and wisdom of God. In the Christian tradition the naming of species is equivalent to the domination and conquest of nature. It can, admittedly, be argued that the Linnaean system soon went well beyond its creator's pious premises when it was adopted by the European scientific community. But in so doing it became even more overtly imperialistic in character as the new science was practised in a purely Western tradition. In the first place the Linnaean naming of species was culturally determined from the outset. In his introduction to *Systema naturae* Linnaeus wrote that a species inspires its own name and that this name proclaims what has been known about it through time.[26] This, of course, meant that 'scientific' naming was based on what a Forsskål, Geoffroy or Ehrenberg discerned as characteristic of a species, and that it reflected the knowledge *they*, through *their* Western scientific tradition, had acquired about a plant or animal—or, as Pratt has it, 'a European discourse about non European worlds'.[27] And finally, to return to Said's argument, the naming (and ordering) of a species by the Danes, French and Germans was an appropriation of knowledge which was ultimately transferred to a European textual tradition. Their collections were punctually shipped to, and described and published in, Copenhagen, Paris and Berlin in order to constitute the collective body of European knowledge.

23. Said, *Orientalism*, p. 334.
24. Said, *Orientalism*, p. 86
25. See also Mary Louise Pratt, *Imperial Eyes: Travel Writing and Transculturation*, London 1992, pp. 24–37, 56–7.
26. 'Methodus, anima Scientiae, indigitat, primo intuitu, quodcunque Corpus naturale, ut hoc corpus dicat proprium suum Nomen, & hoc nomen quaecunque de nominato corpore beneficio seculi innotuere, ut sic in summa confusione rerum apparenti, summus conspiciatur Naturae ordo.' Carolus Linnaeus, *Systema naturae,* Vienna 1767, p. 13.
27. Pratt, *Imperial Eyes*, p. 34.

This book in the series Studies in The Arcadian Library deals with these prominent post-Linnaean collectors. It follows the vicissitudes of the naturalists of the Danish, French and Prussian expeditions on their adventurous quests for new plants and animals in the Near and Middle East. The naturalists' scientific publications, their travel diaries and their private letters have been used to reconstruct the history of these expeditions, and modern taxonomical literature has been consulted to gauge the scientific importance of their discoveries.

Historians have done little work on the botanical and zoological results of the first two major expeditions, and no study has yet traced the vicissitudes of the naturalists on the German expedition or examined their scientific achievements in depth. The three expeditions are here brought together for the first time in a single study. I show that, although the Danish, French and German expeditions were in direct competition with one another as a question of national and personal prestige, the expeditions were at the same time European joint ventures to catalogue the animal and plant species in the Arab World.

For the Danish expedition I have relied heavily on Carsten Niebuhr's accounts, Forsskål's scientific publications and private travel diary, as well as on Michaelis' *Fragen an eine Gesellschaft gelehrter Männer* (1762), which provided a challenge and a guide to the members of the expedition. To reconstruct the French enterprise in Egypt I have mainly looked at the natural history volumes of the *Description*, the papers presented at the Institut d'Égypte in Cairo, and at Geoffroy's private letters. In the scientific descriptions of *Symbolae physicae* little concerning the Prussian expedition emerges, and Ehrenberg's travelogue *Naturgeschichtliche Reisen* (1828) only covers the first few months of their journey. However, Ehrenberg's and Hemprich's private correspondence of 1820 to 1826 sheds abundant light on their activities and adventures. These were not only characterized by disappointments and tragedy, but also by unbounded enthusiasm at new finds.

This book concludes with a historical and scientific account of two further late nineteenth-century scientific expeditions to the island of Soqotra about 500 miles south-east of Aden. It was visited in 1880 by an English expedition led by Isaac Bayley Balfour, and in 1881 by a German expedition led by Georg Schweinfurth. The botanical results of these two expeditions were published by Balfour in his *Botany of Socotra*, a work which described for the first time the island's numerous endemic species. Not much is known about the expeditions themselves, except for a short account by Balfour and the notes in a recently found travel diary of the German expedition, which is now in The Arcadian Library. These are used for my account of the English and German exploration of the Soqotran flora and fauna. Historians have paid very little attention to the Soqotra expeditions, and many aspects of these late nineteenth-century enterprises are here explored for the first time.

The core of my research has been based on the important collections of The Arcadian Library in London. The library not only contains rare European books testifying to five centuries of relations between Europe and the Arab world, but also holds a fine collection of important scientific publications, including copies of many of the above-mentioned works. Most of the illustrations reproduced in this volume are from the collections of The Arcadian Library.

CHAPTER I

PEHR FORSSKÅL AND THE DANISH EXPEDITION OF 1761–1767

Copenhagen and Göttingen

Georg Wilhelm Baurenfeind, engraving of view of Rosetta (detail).
Carsten Niebuhr, *Reisebeschreibung nach Arabien und andern umliegenden Ländern*, Copenhagen 1774, plate VI [following page 56]
The Arcadian Library, 12540

In the late 1750s an ambitious project for a Danish scientific expedition to the Yemen took shape. It was to become the first full-scale scientific expedition to the Near East. There were many factors that made Denmark open to such an enterprise. During the late 1740s and early 1750s King Frederick V (1746–66) had become famous for his programmes favouring and stimulating the arts and furthering the sciences. He founded schools and military academies in both Denmark and Norway, then still under Danish rule, and established academies of printing and sculpture in the capital, Copenhagen. His father too, Christian VI (1730–46), had made important contributions to the nation's interest in learning and culture. In 1742 Christian had founded the Royal Danish Academy of Sciences and Letters. However, as he was driven by religious ideals, he committed himself to the cause of a national church influenced by German Pietism, and his encouragement of science and art remained subservient to his religion. Although Frederick inherited his father's interest in learning and religion, he freed it from the pietistic restraints that had dominated it during the 1730s and early 1740s. In this new climate of royal patronage, the arts and sciences prospered. Moreover, as countries such as France, Austria, Saxony, Sweden, Russia, Great Britain, Hanover and Prussia were hampered by their involvement in the Seven Years' War, Frederick, through the able manoeuvring of his foreign minister Johann Hartwig Ernst von Bernstorff, managed, in 1756, to steer clear of the conflict, and preserve a marked neutrality. All this made Denmark a suitable sponsor for a scientific expedition to the Near East. No doubt it was with such considerations in mind that the Hebrew scholar and theologian Johann David Michaelis addressed a letter to Frederick V saying that a study of the Yemen might yield fruitful new knowledge for the history of Christianity.

Michaelis had been instructed in oriental languages at home by his father before he enrolled as a student at Halle University, where his father taught Syriac. The religious attitude that he had derived from Halle changed when he went to the Netherlands and England in 1741–2. In the field of philology he was much impressed by the new methods of the Dutch philologist Albert Schultens at Leiden, and he adopted a form of deistic empiricism. He came to see scripture as a text which could, and should, be subjected to the independent historical and linguistic exercise of judgement ('criticism'). Although the University of Halle might initially have seemed a natural place to return to, by now it was intellectually far too narrow for his modern views on historical scholarship. In 1745 he was appointed *Privatdozent* at the Georg August University of Göttingen in the nearby Electorate of Hanover. Although the university had been founded as recently as 1734 (and opened in 1737), it had rapidly become an important centre of learning which assumed a leading position in Europe. Apart from Michaelis, it counted amongst its professors during the late eighteenth century such famous scholars as the Swiss anatomist and physiologist Albrecht von Haller, the physicist Georg Christoph Lichtenberg, and the historian August Ludwig von Schlözer, who is reputed to have been the originator of the motto 'Extra Gottingam non est vita, si est vita non est ita!' ('There is no life outside Göttingen; if there is, it is not really life').

By the mid-1750s Michaelis, who was almost immediately appointed *Professor extraordinarius*, and subsequently, in 1750, *Professor ordinarius*, was considered as one of the leading philologists and orientalists in Europe. In later life his fame was such that men like Benjamin Franklin, Alexander von Humboldt, Gotthold Ephraim Lessing and Johann Wolfgang von Goethe came to stay with him at his house in Göttingen.[1] Although his philological work is nowadays considered inferior to that of his colleagues Albert Schultens at Leiden and Johann Jakob Reiske at Leipzig, his emphasis on the study of Hebrew history in the larger context of ancient Near Eastern history laid the groundwork for the higher criticism that was to change biblical studies profoundly in the nineteenth century. Michaelis maintained that biblical scholarship could no longer be based on textual criticism alone, as had been done in previous centuries, but that it should be cast as a new comprehensive science that included anthropology, archaeology, ethnology and natural science.[2] He ultimately believed that historical investigation would prove Christianity to be true, but at the same time he was prepared to accept conclusions that showed traditional beliefs to be false. In a modern assessment J.C. O'Neill concludes that 'Michaelis broke clearly and thoroughly from dogmatic attempts to read all the evidence according to a set theory', and that in so doing 'he founded and established modern critical and historical study of the Bible'.[3] Critics have been less kind about his contribution to the study of Arabic. The German Arabist Johann Fück describes Michaelis as a gifted man without originality, who merely popularized and summarized the findings of other scholars, and who studied Arabic in the traditional context of *philologia sacra*.[4]

Michaelis' interests were by no means limited to the philology of oriental languages. Apart from historical studies of the Old and New Testaments, he also had a strong curiosity about geography and natural history, as well as a passion for modern literature. He used to describe himself as a systematic student of world knowledge ('ordentlicher Lehrer der Weltweisheit').[5] A curious example of his scholarly activities is his German translation of Samuel Richardson's novel *Clarissa*, as well as his *Spicilegium geographiae hebraeorum* (1769–80). It was such wide-ranging interests that made Michaelis write in May 1756 to the foreign minister of Denmark with his plans for a scientific expedition to the Near East, and in particular to the south-west of the Arabian peninsula. In the Yemen, Michaelis explained in his letter to Bernstorff, a philologist who also possessed a firm grasp of the natural sciences might discover much about the linguistic and ethnological background of the Bible. Although many

1. See Ulrich Hübner, 'Johann David Michaelis und die Arabien-Expedition 1761–1787', in Josef Wiesehöfer and Stephan Conermann, *Carsten Niebuhr (1733–1815) und seine Zeit*, Stuttgart 2002, p. 370.
2. Cf. Jonathan Sheenan, *The Enlightenment Bible: Translation, Scholarship, Culture*, Princeton and Oxford 2005, pp. 182–217.
3. J.C. O'Neill, *The Bible's Authority: A Gallery of Thinkers from Lessing to Bultmann*, Edinburgh 1991, p. 38.
4. Johann Fück, *Die arabischen Studien in Europa bis in den Anfang des 20. Jahrhunderts*, Leipzig 1955, p. 119.
5. See Michaelis' inaugural lecture 1750; cf. Hübner, 'Johann David Michaelis und die Arabien-Expedition 1761–1767', p. 367.

travel works on Palestine and Egypt had been written, only very few descriptions of Arabia Felix, as the Yemen had been called from antiquity, were available to the modern scholar. He insisted that an expedition to the Yemen would certainly reveal unknown aspects of nature, while its ancient history would shed new light on many passages of the Old Testament. Its dialect too, he continued, different from the familiar Western Arabic, would provide new insights for the interpretation of the Hebrew text of the Bible, especially since Western Arabic had so far been of the greatest assistance in explaining Hebrew. Michaelis hoped that what had disappeared in one dialect might well have been preserved in another, as, for example, the names of many plants and minerals.[6]

Michaelis urged, moreover, that with such an expedition Denmark could earn much honour as an enlightened kingdom: 'I made so bold as to hope that if there were a monarch anywhere who would give science such an auspicious gift, this would be the northern Augustus of our times.'[7] This point carried weight with the susceptible Danish king, all the more so since the competing Royal *Swedish* Academy of Science had recently made impressive advances in the field under the leadership of Linnaeus. Michaelis, therefore, found a willing ear in Bernstorff, who, apart from being an able politician, also had a strong interest in the arts and sciences. In the ensuing correspondence the project soon gained in dimension. Michaelis had initially proposed tentatively that, as Denmark regularly sent missionaries to Tranquebar (since 1616 an important Danish trade colony in southern India), some of these be properly trained in the basics of oriental languages and natural science so that they could try to find answers to a series of scholarly questions as they passed Southern Arabian settlements on the trade route to India. Michaelis emphasized that this was a relatively cheap way to gather invaluable material for the study of the Old Testament. Apart from linguistic research, he was thinking of the geography, animals and plants of the area, as well as a study of the tides of the Red Sea to shed light on the miraculous Exodus of the Jews from Egypt. The only expense of any consideration would be the costs involved in the preparation of the missionaries. He stressed that they should have a good working knowledge of Arabic before leaving, as he thought it was impossible to learn the grammar and spelling of the language correctly once on the spot. This was essential for a scientific expedition to succeed, he urged: 'How much depends in geography and natural history on how names are written and pronounced locally.' The mere transcription of Arabic had no scientific value. 'How much confusion and obscurity', he argued, 'has been caused when such

6. 'Dieß Land ist noch an uns unbekannten Geschenken der Natur reich: seine Geschichte steiget bis in die alleraltesten Zeiten hinauf: seine Dialect ist von dem uns bekannten westlichen Arabischen noch verschieden, und da dieß Arabische, welches wir kennen, bisher das sicherste Hülfsmittel zur Erklärung des Hebräischen gewesen ist, was für Licht müßen wir denn billig für das allerwichtigste Buch des Altertums, für die Bibel, erwarten, wenn wir die östliche Dialect Arabiens so gut kennen lernten, als die westliche? was in der einen Mundart untergegangen ist, wird vielleicht in der andern übrig seyn, z. E. die Namen von manchen Pflanzen oder Edelgesteinen.' Johann David Michaelis, *Fragen an eine Gesellschaft gelehrter Männer, die auf Befehl Ihro Majestät des Königes von Dänemark nach Arabien reisen*, Frankfurt 1762, 'Vorrede,' [p. 9]; see also Michaelis' long letter of 1756 to Bernstorff in Johann David Michaelis, *Literarischer Briefwechsel*, ed. Joh. Gottlieb Buhle, Leipzig 1794, pp. 299–324.

7. '[I]ch wagte es zu hoffen, wenn irgend ein Monarch den Wissenschaften ein so erwünschtes Geschenk machen würde, so würde es der nordische August unsere Zeiten seyn.' Michaelis, *Fragen*, 'Vorrede,' [p. 10].

names are wrongly written down!'[8] Michaelis also underscored the fact that the missionaries should be given a list of questions, drawn up by European scholars, to which they should try to find answers. So many new impressions would crowd their minds that, without realizing it, they could easily overlook a fact which a scholar back in Europe would like to see described.[9] Without such preparation the value of the expedition would be much diminished. In fact it was such shortcomings that had made most previous journeys to the Near East scientifically less successful than they might have been.[10] Michaelis, therefore, proposed that two students should receive a three-year scholarship at Göttingen, at the expense of the Danish government. They would then go to Copenhagen to instruct the missionaries. He also mentioned that he had two students at Göttingen just then who would be suitable for such a task, both subjects of King Frederick V: Jan Heinrich Strøm, a Norwegian theology student, and Friedrich Christian von Haven, a Dane.

The Danish reaction was, on the whole, enthusiastic, but before the king could approve, he requested a more detailed travel programme. Encouraged by such an eager interest, Michaelis ventured to modify his plan and explained that, after all, he did not want to employ a missionary for such important scholarly work. He thought it well worth spending more money to make the enterprise a success, and to send out a professional linguist and scientist on an expedition to study the language, geography and natural history of southern Arabia, and not depend on the missionaries who merely passed through the region on their way to Tranquebar. So Michaelis now had to choose between his two students, and his recommendation fell on Strøm. Frederick V accepted the new plan, and Michaelis' student was appointed. The Danish government, or rather Bernstorff, now entered so fully into the enterprise that they decided also to appoint a botanist to accompany Strøm. The Norwegian linguist, however, soon changed his mind, and opted out of the project. Michaelis' choice now was Von Haven, who had let the professor know that he envied Strøm's appointment from the beginning. Michaelis assured Bernstorff of Von Haven's qualities and that he had 'the greatest desire to undertake this expedition, and is thus in possession of a quality that can be acquired neither by instruction nor for money'.[11] The Danish government agreed.

In his letter of recommendation to Bernstorff Michaelis had also claimed that, although principally a philologist, Von Haven was 'fairly familiar with the necessary auxiliary disciplines, including botany'. But, if the Danish government had initially decided to add a botanist to the expedition because Strøm was a man with hardly any botanical knowledge, now that they had a linguist who apparently did feel confident in botany they saw no reason to change their mind about sending a professional botanist as well. This shows that at an early stage the Danish government saw the

8. 'Wie viel kommt in der Geographie und Naturgeschichte auf Namen an, wie sie an Ort und Stelle geschrieben und ausgesprochen werden? und was für Verwirrung oder Dunkelheit enstehet daraus, wenn diese Namen unrichtig geschrieben sind?' Michaelis, *Fragen*, 'Vorrede,' [pp. 2–3].

9. 'Sie sehen unzählige Dinge vor Augen, auf die sie nicht merken; allein sie würden ihnen einen Blick gönnen, wenn sie wüßten, daß gerade dieß die Sache sey, darnach ein Gelehrter in Europa fraget, und die er zu Aufklärung einer Dunkelheit gebrauchen kann. Soll daher ihre Reise recht nützlich werden, so muß man ihnen Fragen vorlegen', Michaelis, *Fragen*, 'Vorrede,' [pp. 5–6].

10. Michaelis, *Fragen*, 'Vorrede,' [p. 8].

11. Thorkild Hansen, *Arabia Felix: The Danish Expedition of 1761–1767*, London 1964, p. 18.

growing importance of the scientific side of the expedition and was willing to extend the original project consistently, and meet the increasing expenses. It was just as well that they insisted on keeping a separate botanist on the expedition, as it soon turned out that Von Haven was not proficient in natural history at all.

With regard to Michaelis' selection of Von Haven for the expedition, Barthold Georg Niebuhr commented decades later in his 'Carsten Niebuhrs Leben' that it came as a surprise that the choice did not fall, as expected, on Johann Jakob Reiske who was widely considered to be unrivalled in his knowledge of Arabic, and he added that this was all the more astonishing seeing that Von Haven's knowledge of Arabic must have been very slight at the time.[12] On the whole, Niebuhr's description of Von Haven's character and qualities was devastating. He wrote, for example, that Von Haven had chosen a profession for which nobody was less suited than he, and that his only thought was to return home. He always talked of a comfortable future after his return, and no discoveries or observations abroad could make him forget the difficulties and discomfort he suffered during his journey. No other member of the expedition felt as much deprived of his comforts as he. A good meal, Niebuhr continued, with a good bottle of wine were for Von Haven life's highest goal, and in Arabia, where the travellers had little to eat and only bad water to drink, his discontent soon turned into despair. Von Haven was by nature lazy and always and everywhere found excuses not to work. Moreover, he felt himself to be superior to the others. After his death, Niebuhr concluded, not the least bit of useful knowledge could be found in his all too slender diary.[13]

Niebuhr's assessment seems to have been the basis for Hansen's negative portrait of the Danish philologist in his account of the expedition in his *Det Lykkelige Arabien* (1962). The lack of further testimonies makes it impossible to decide whether Von Haven was indeed as dislikeable and lazy a character as Niebuhr maintained. His scientific reputation, on the other hand, has recently been amply redeemed by modern historians. Rasmussen has stressed the importance of the manuscripts Von Haven purchased for the Royal Collection, and Egon Keck has confirmed that Von Haven's choice of manuscripts proves his competence as a philologist.[14] This assessment goes some way to justifying Michaelis' recommendation to the King of

12. '[W]er hätte nicht erwarten sollen, daß dieser [Michaelis] den Mann vorgeschlagen haben würde, der unter allen seinen Zeitgenossen ohne Nebenbuhler im Ruhm der arabischen Philologie stand,' while Von Haven's 'Kenntnisse damals ganz schülermäßig gewesen sein müssen'. Niebuhr, Barthold Georg, 'Carsten Niebuhrs Leben', in *Carsten Niebuhr, Reisebeschreibung nach Arabien und andern umliegenden Ländern*, ed. Wolfgang Stammler, Zürich 1992, pp. 890–1.

13. 'Er hatte aber überhaupt einen Beruf gewählt, zu dem niemand weniger paßte. Sein einziger Gedanke war zurückzukehren, sein liebstes Gespräch von der gemächlichen Zukunft, die er sich alsdann versprach; kein Trieb nach Entdeckungen und Beobachtungen machte ihn die Mühseligkeiten und Entbehrungen der Reise vergessen, und niemand entbehrte und vermißte soviel wie er. Ein leckerer Tisch und guter Wein waren für ihn der höchste Reiz des Lebens, und in Arabien, wo die Reisenden nur spärliche Kost für den Hunger und schlechtes Wasser fanden, stieg sein Unmut zu einer Verzweiflung, welche die beiden Berufenen unter den Reisegefährten häufig belustigte, manchmal aber auch empörte. Er war von Natur faul und fand sich unter jenem Himmelsstrich völlig entschuldigt, nichts zu arbeiten Übrigens zeigte er sich manchmal hochmütig und dünkelvoll [...]. Es hat sich nach seinem Tode auch nicht das allergeringste Brauchbare in seinen magern Journalen gefunden.' Barthold Georg Niebuhr, 'Carsten Niebuhrs Leben', p. 894.

14. Egon Keck, 'De hebraiske manuskripter og Kennicott-biblen', in Stig Rasmussen, ed., *Den Arabiske Rejse 1761–1767: En dansk ekspedition set i videnskabshistorisk perspektiv*, Copenhagen 1990.

Denmark that he could not think of a philologist for the expedition from whom more could be expected than Von Haven.[15]

It is not clear on what evidence Niebuhr based the affirmation that Reiske's name was expected to be put forward by Michaelis as the ideal philologist for the expedition. Of course, Carsten Niebuhr was clearly prejudiced against Von Haven, a sentiment that was shared by all the other members of the expedition. He might, in retrospect, have passed on a biased interpretation of events to his son to show how erroneous was the choice of Von Haven. Still, Niebuhr's hypothesis that Reiske was at the time generally considered a suitable candidate for the post gains some plausibility when one recollects that it was during the second half of 1756, when the selection of an able philologist was being discussed, that Reiske, who, as 'was known in all Germany',[16] had been reduced to dire poverty, turned to Michaelis for help. Reiske and Michaelis had been schoolfellows once, and Reiske hoped Michaelis could do something for him at the University of Göttingen. Johann Fück comments that Reiske had not reckoned with Michaelis' cool, calculating, and selfish nature.[17] Indeed, an obvious way of helping an old school companion would have been to propose his candidacy for the Danish expedition. In reality, however, Reiske, who had repeatedly argued that philology had to be separated from theology, would have been a most unlikely person to investigate Michaelis' questions, which were above all meant to give a new impetus to Bible exegesis. Nor was Reiske the kind of person who was willing to work for others—he had refused a good offer of work as d'Orville's amanuensis in Amsterdam in 1738[18]—and he was therefore incompatible with Michaelis' plans for the Danish expedition. Von Haven seemed a more reliable choice at the time.

Towards the end of 1756, with the appointment of Von Haven and the professional botanist whose name is not known, the ground-plan for the expedition to Arabia had been accepted by royal Danish mandate. But another five years would pass before the expedition set sail. A series of delays and changes made continuous postponement necessary. First, Von Haven, who had been given two years to prepare himself, soon made it clear that he could not handle the geographical surveying alone and suggested that a trained mathematician join the expedition. Although the Danish government did not object to the further expenses this would involve, it proved difficult to find a suitable candidate who was willing to embark on the Arabian adventure. More than a year passed, and neither Michaelis in Göttingen nor Bernstorff in Copenhagen was able to come up with anybody. When time ran out half-way through 1758, Michaelis proposed a German, Carsten Niebuhr, a promising young student of humble origins who had no ties to keep him in Germany. He was to be the sole survivor of the expedition and the one who was to chronicle in detail the vicissitudes of the Arabian journey. As soon as the problem of finding a mathematician had been solved, a second problem arose: the unnamed botanist Bernstorff had chosen for the expedition was no longer willing to go. The Danish foreign minister once more consulted Michaelis, who announced only months later, on 1 January 1759, that

15. Michaelis, *Fragen*, 'Vorrede,' [p. 12].

16. Barthold Georg Niebuhr, 'Carsten Niebuhrs Leben', p. 890.

17. Fück, *Die Arabischen Studien in Europa*, p. 119.

18. Fück, *Die Arabischen Studien in Europa*, p. 109.

At last I have found somebody who is eminently suited to act as botanist on the Arabian Journey, somebody who indeed almost exceeds my expectations, a former student of mine, in fact, by the name of Forsskål, Swedish by birth. Before he ever came to me, he had studied Arabic and he has also distinguished himself in botany—and indeed in natural history in general. He need not undergo any preparatory training at all.

To moderate this glowing description somewhat, Michaelis warned that 'His one fault is an almost excessive scepticism, but this makes him if possible even more suited for the expedition.'[19] But, as he explained a few years later, Forsskål's mild criticism in his doctoral thesis (*Dubia de principiis philosophiae recentioris*) of some of Michaelis' theorems had convinced him of the young man's suitability for the enterprise: 'I knew for sure that he would never believe anything without reason, and that he loved the truth; his philosophic dispute with me was a guarantee [...] that he would never ever believe anything true if he had not heard and seen it with his own ears and eyes.'[20] Bernstorff again accepted Michaelis' candidate unreservedly.

Pehr Forsskål was indeed a brilliant young scholar. Born in Helsinki, which was then part of the Kingdom of Sweden, on 11 January 1732, he moved with his father, a Lutheran pastor, to Tegelmora in Sweden at the age of nine. His father's parish lay in the vicinity of the University of Uppsala, where the boy enrolled as a student of theology at the tender age of ten. His studies included philosophy and oriental languages, which were deemed important auxiliaries in the study of theology. When Professor Aurivillius introduced the study of Arabic at the university, Forsskål effortlessly acquainted himself with the basic principles of this language too. Although his early interests at Uppsala were mainly modern philosophy, he also fell under the spell of Linnaeus, whose lectures he attended. He started to describe birds and became absorbed in observing the metamorphosis of insects.[21] His studies in botany under Linnaeus had made him a keen disciple of the new system of classification.

Ever since the publication of *Systema naturae* in 1735 Linnaeus had been working on a new taxonomic system which abandoned the Aristotelean method of classification that had been used for over two thousand years. In 1751 he brought out his influential *Philosophia botanica* in which he affirmed the possibility of drawing up a natural system of classification for God's immutable creation. He drew up a hierarchy of classes, orders, genera and species, and on the basis of this created functional keys to identify plants and animals. He stressed above all the pivotal significance of the reproductive organs in the classification of plants, and therefore worked on detailed analyses of their flowering parts. Using the descriptions of these flower parts, which had been neglected by botanists till then, he created a binomial taxonomic system, which he worked out in *Species plantarum* (1753) and in the tenth

19. Hansen, *Arabia Felix*, p. 19.

20. 'Ueberhaupt wußte ich, daß er nicht leicht glaubte, ohne dazu durch Gründe gezwungen zu seyn, und daß er die Wahrheit liebete: und sein philosophischer Widerspruch gegen mich war mir Bürge davor, daß er [...] niemals im Orient etwas zu sehen oder zu hören glauben würde, das er nicht wirklich sähe und hörte.' Michaelis, *Fragen*, 'Vorrede,' [pp. 15–16].

21. Pehr Forsskål, *Flora Aegyptiaco-Arabica: Sive descriptiones plantarum, quas per Aegyptum Inferiorem et Arabiam Felicem detexit, illustravit Petrus Forsskål. Post mortem auctoris editit Carsten Niebuhr. Accedit tabula Arabiae Felicis geographico-botanica*, Copenhagen 1775, pp. 31–2.

edition of his *Systema naturae* (1758). These publications are recognized today as the starting point of modern taxonomy.

The years in which Linnaeus consolidated his system of botanical and zoological description at Uppsala coincided with the period when Forsskål was studying there. By 1751 Forsskål had distinguished himself at the University and in recognition had been awarded a five-year scholarship. What must have aroused the young Forsskål's enthusiasm about his countryman's new taxonomy was the fact that the keys and descriptions in his books made the identification of plants relatively easy. It was not until the beginning of the nineteenth century, however, that Linnaeus' binomial system became generally accepted by botanists. During the 1750s the Danish naturalists at Copenhagen were little inclined to accept Linnaeus' innovations in classification. Much of the value of Forsskål's botanical and zoological work done during the journey to the Yemen lies in the fact that it was carried out in the framework of the Linnaean system. Still, his loyalty to the Swedes was to cause some bitter controversy before the expedition could sail.

Danes and Swedes

Having settled on a philologist, a botanist/zoologist and a mathematician/astronomer, the Danish government proceeded with the final preparations for the long-awaited departure of the expedition. Yet, it was not until January 1761 that the group of young scholars left Copenhagen. In the meantime the Danish government, on the advice of professors Peder Ascanius and Georg Christian Oeder at Copenhagen, had decided to add a physician and an artist to the team. Seeing that the proposal came from two Danish professors, the choice of the candidates was theirs, and Michaelis was not consulted this time. The Danish government also decided to appoint a general servant to look after the daily needs of the expedition. For the servant the choice fell on a Swedish dragoon by the name of Berggren, a practical, sturdy man who had fought in the war against Prussia, while Ascanius and Oeder singled out the German Georg Wilhelm Baurenfeind as a capable engraver who could make invaluable illustrations. These appointments were not controversial. With the physician it was otherwise. When Forsskål arrived in Copenhagen on 1 October 1760 he drafted a letter to Bernstorff in which he asked, amongst other things, for the appointment of an assistant to help him with his botanical work. For this job he strongly recommended Johann Peter Falck, another Swede who had studied under Linnaeus in Uppsala. Ascanius and Oeder, however, had in the meantime proposed Christian Carl Kramer, a Dane who had studied medicine and natural history at the University of Copenhagen under Professor Christian Gottlieb Kratzenstein. The government's intention to appoint Kramer as physician and assistant natural historian infuriated Forsskål, partly because he felt accused of Swedish nationalism in his choice of Falck. For Forsskål, however, the point was that he thought Falck highly superior in training to Kramer, not least because the Swede had been trained under Linnaeus. As Kramer needed to be instructed by him in most branches of natural history, he complained to Bernstorff, he could be of little assistance in the scientific work of the expedition.

If Forsskål had left it at this he might still have persuaded the Danish government to add Falck to the company. However, his reasons for maintaining that Kramer needed to be instructed in natural history by him first must have found little sympathy in Denmark. Studies in natural history, he added in his letter to Bernstorff, were of a much higher standard in Uppsala than in Copenhagen, where natural science was only in its infancy. Besides, in this affront to the Danish state, Forsskål profoundly offended Kratzenstein, who, in a letter to Bernstorff, ascribed Forsskål's preference for Falck as proof of his arrogant Swedish nationalism. Kramer was consequently appointed. Kratzenstein reports that soon after this decision he was visited by Forsskål in person, who behaved in a hostile and ill-tempered way. He told him that he, Kratzenstein, was much inferior to Linnaeus, that compared to the latter he was a 'mere dwarf'. Not surprisingly, Kratzenstein could scarcely refrain from getting up and striking the young man in the face. The way Forsskål handled his opposition to the appointment of Kramer has been seen as an example of his irascible temperament.[22]

What is important to stress about Forsskål's stand in this matter, however, is his serious commitment to the scientific task that lay ahead of him. This had much to do with his pride in being one of Linnaeus' disciples, and his conviction that he would perform pioneering research in the field his master had revolutionized. To his diary he confided—albeit with some awareness that it would one day be read in Denmark and Sweden—that he felt honoured, as well as lucky, to have been entrusted with the study of natural history during the expedition, a field to 'whose principles I myself had been introduced by my venerable master, the great Linnaeus'.[23] In a letter to the master himself, on the other hand, he lamented that it was 'very difficult not to be offended with them here [in Copenhagen] and be coupled to an unreliable fellow'.[24] Two months later he admitted in another letter to Linnaeus that Kramer's appointment spoiled the journey for him. Try as he might, there could be no understanding between them, so each one would have to work for himself.[25] His reaction to Kramer's professional inadequacy, moreover, meant that he 'had to endure a lot of intrigue, which has even reached the point where they have tried to accuse me of ill-feelings towards the Danish nation'.[26] Forsskål maintained that he did show some understanding for the decision of the Danish government. But the point was not the man's nationality, but whether he was a capable natural historian or not. It was quite understandable, he explained to Linnaeus, that Kratzenstein, Ascanius and Oeder wanted another *Dane* on a *Danish* expedition, so long as science did not suffer from such a choice and as long as he had not brought Falck to Copenhagen for the very purpose.[27] As it stands the last reason is ambiguous.

22. Hansen, *Arabia Felix*, p. 41.

23. '[...] hvars grunder jag sjelf undfått af min vördade Läromästare, den store Linnée', Pehr Forsskål, *Resa till Lyckliga Arabien: Petrus Forsskåls Dagbok 1761–1763*, edited with introduction by Arvid Hj. Uggla [from now on *Dagbok*], Uppsala 1950, p. 3.

24. 'Jag får swår sysla att ej stötas med dem här eller ej bindas wid en opålitelig kamrat.' Forsskål to Linnaeus, 4/10/1760, in Th. M. Fries, J. M. Hulth and A. Hj. Uggla, *Bref och skrifvelser af och till Carl von Linné. Första afdelningen*, vol. vi, *Bref till och från Svenska enskilda personer*, Stockholm 1912, p. 126.

25. Forsskål to Linnaeus, 14/12/1760, in Fries, Hulth and Uggla, *Bref och skrifvelser,* pp. 128–29.

26. 'Jag har rönt för den saken många intriguer, som ock gådt så långt, att man sökt anskrifwa mig som illasint mot danska nationen.' Forsskål to Linnaeus, 14/12/1760, in Fries, Hulth and Uggla, *Bref och skrifvelser,* p. 129.

27. Forsskål to Linnaeus, 4/10/1760, in Fries, Hulth and Uggla, *Bref och skrifvelser*, p. 127.

If it means that he, Forsskål, had suggested a highly qualified naturalist, then it is in line with his argument that science should not suffer on account of the choice of the scientist. It sounds, however, as if it conceals an underlying personal grudge against the Danish government after he himself had gratuitously brought a Swedish candidate to Copenhagen.

Some assessment of Forsskål's partiality in the name of science might lie in the comparison of the two candidates. Forsskål wrote disparagingly about Kramer's achievements in natural science. He is but a a poor ('eländig') doctor of medicine, he maintained, who in his doctoral dissertation *Specimen insectologiae danicae* had not described a single new species and had copied everything from Friedrich Christian Lesser's *Insectotheologia* (1738).[28] Admittedly Kramer's work on insects was hardly an original contribution to natural science. Nor is there any proof of significant work done by him during the Arabian expedition till his premature death in Bombay in 1763. But to say that he was an 'eländig' scientist is an exaggeration. It is in comparison with Falck's later achievements that Kramer loses stature as a scientist. Hansen describes him as a decent and sociable, but *lazy* man.[29] Falck, on the other hand, although he had not published anything significant by 1760, was seen as a most promising naturalist by Linnaeus. After he failed to be appointed to the Danish expedition, he was sent by Linnaeus to St Petersburg, where he was nominated curator of the botanical gardens. In 1768 he was chosen to lead a scientific expedition to eastern Russia and Siberia, and, although he was unable to publish his finds—he committed suicide in Kazan in 1774 before returning from the expedition—his explorations in Russia formed a genuine contribution to botany. The genus *Falckia* was named after him.

History has thus proved Forsskål right in praising Falck's qualities and disparaging Kramer's. Still, what undoubtedly conditioned him in his preference for Falck from the very beginning was, as he explained in a letter to Bernstorff, the fact that he had been instructed by the great Linnaeus. That Forsskål bore no personal grudge against a man he did not know is also borne out by a comment in a letter written from Istanbul on 1 August 1761, when Forsskål had known the 'decent and sociable' Kramer for a full seven months. He still grumbled to Linnaeus that science should not have been allowed to suffer simply because there were unreasonable people in Denmark.[30]

The King's 'Instruction'

On 15 December 1760 Bernstorff drafted a document for the members of the expedition in which their individual tasks were specified. This was the king's 'Instruction', based on the advice of a group of European scholars.[31] In the early stages

28. Forsskål to Linnaeus, 1/1/1761, in Fries, Hulth and Uggla, *Bref och skrifvelser,* pp. 131–2; his own dissertation was presented to the University of Copenhagen on 29 December 1760, only a few days before the expedition set sail.
29. Hansen, *Arabia Felix*, p. 39.
30. Forsskål to Linnaeus, 1/8/1761, in Fries, Hulth and Uggla, *Bref och skrifvelser,* p. 143.
0031. The original of the Instruction is no longer extant (see Stig Rasmussen, '"Niebuhriana" in Kopenhagen', in Josef Wiesehöfer and Stephan Conermann, *Carsten Niebuhr (1733–1815) und seine Zeit*, Stuttgart 2002, p. 43.

of the project Michaelis had published a request for advice in a series of newspapers, asking what questions scholars would like to see answered about the Yemen and its language. Although Michaelis felt disappointed by the limited number of reactions, those who showed an interest included De Navarre of Amsterdam, Thiéry of Paris, and Collet of London.

The Instruction, a manuscript copy of which was given to each member of the expedition, but which was printed in part for the first time in 1762 with the more specific questions for the 'company of learned men who on the order of His Majesty the King of Denmark travelled to Arabia', emphasized the necessary collective spirit of the enterprise. The first article decreed that the travellers should travel together ('saemmtlich mit einander') to Arabia Felix and that on reaching their destination, as well as during the outgoing sea voyage, they collect as much scientific material and information as possible. They were supposed to travel through Egypt to the Sinai, then go down the Red Sea to Mocha,[32] and from there into the interior of the Yemen, where they should collect information for two or three years.

Another cardinal point of the Instruction was that all the members of the expedition should not entrust anything to memory, but instead each keep a diary in which to record, possibly at the end of the day, what had been observed. The annotations had to be complete and clear so as to be fully intelligible to any other person. This was in case death prevented the traveller from explaining his own words personally. Copies of the diaries should be sent from time to time to Count Moltke, but not until they had been read out to other members of the expedition. This was not to check on discrepancies between the diaries, but in order to see whether the content itself was comprehensible.[33]

The members of the expedition were expected to behave courteously to the Arabs, not to argue about religion, and not to intervene whenever anything conflicted with their European standards.[34] This policy of non-interference with Arab culture did not apply to Kramer's medical knowledge, which he was allowed to put to the service of the Arabs whenever the occasion arose—an apparently humanitarian recommendation which also had a useful side. It would, the Instruction specified, earn the expedition the confidence of the native population and thus 'open new ways to penetrate Arabia further'.[35]

Marseilles and Malta

With the appointment of Kramer, the expedition was ready to sail. On 4 January 1761 the members left on the *Grönland*, a 50-cannon war vessel under the

The following discussion of it is based on the extracts Michaelis gives in his *Fragen* (1762).00

32. Point 3 of the Instruction reads 'über das Rote Meer nach Mecca'; if Mecca had been the real destination of the expedition, one would have expected Jiddah here. Mecca is, of course, not situated on the Red Sea. Moreover, as the members had no intention of visiting Mecca, even if they stayed for a few weeks at Jiddah in October 1762, it is likely that we here have one of the many 'Druckfehler' (misprints) of which Michaelis complains in his postscript to the 'Instruction' ([p. 37]), and that Mecca is a mistake for Mocha.

33. 'Instruction' nn. 8–9.

34. 'Instruction' n. 10.

35. 'Instruction' n. 26.

command of Captain Heinrich Fischer (Fisker) which was to sail from Copenhagen to Marseilles where a Danish convoy lay waiting to be escorted through the Mediterranean to Istanbul. The weather, however, was not propitious. First, days of calm made it impossible to sail, and only by moving sluggishly on the currents in the straits between Denmark and Sweden did they reach Helsingör, the Danish port on the Kattegat 20 miles north of Copenhagen. There they had to wait for wind for another week. Once on the Kattegat a gale arose which made it necessary to return to Helsingör, and they had to wait a further week for favourable winds. During the last week of January the ship finally managed to set sail again and leave the Kattegat behind. In the Skagerrak, however, the *Grönland* was once more struck by a gale and a sailor fell overboard in the 'tremendous seas' and could not be rescued in the dark. Off the Norwegian coast pilots tried to usher the ship into the quiet of a harbour but all to no avail. Fischer was ultimately forced to turn back to Helsingör. They reached it after twenty-four hours of tossing before the wind and lost another sailor.

At this stage Von Haven was deeply discouraged and begged to be allowed to leave the ship. He complained in a letter to Moltke that his life had been one continuous sense of nausea ever since they set sail and he desperately asked whether he was going to be sacrificed to the sea and the terrible winter before he could 'pluck the fruits of glory'.[36] A voyage over the rough northern seas clearly did not agree with Von Haven, and Moltke wisely accepted his request that he travel overland to Marseilles, and meet the rest of the crew there. Forsskål, on the other hand, bore the suffering on board the *Grönland* stoically. Of his travelling companion's desertion he wrote disparagingly that seasickness ('Sjö-sjukdom') is more a matter of discomfort than illness and to be overcome by getting used to it. This, he explained, had nothing to do with the much-feared 'sailor's disease' ('Sjömans-sjukan') accompanied by fevers, coughing, pressure under the heart, headaches, and often diarrhoea.[37] He also added that the latter disease had been described with much learning and experience by the Swedish physician Bierchen.[38] The placing of this piece of *Swedish* wisdom alongside the contemptible reaction of the *Danish* Von Haven is conspicuous. After the unfortunate quarrel over the second naturalist it would seem as if Forsskål could not refrain from denigrating any Dane of learning and from praising all Swedish progress in science. Forsskål's remarks about seasickness gain even more significance when one recalls that, according to royal request, he was to collect zoological and botanical specimens during the voyage, while Kramer had been specifically told to concern himself with the study of diseases peculiar to the region they were going to explore. As for the unpropitious beginning of the expedition, Forsskål jotted down in his diary that the seasons and the winds could not have been personally against them, and that that is why they acquiesced to a sea voyage, where hardly any danger was absent.[39] In Forsskål's typically eighteenth-century rationalist world view there was no room for superstition or belief in Providence.

While Fischer was trying in vain to manoeuvre the *Grönland* out of the Kattegat and Skagerrak, Forsskål was not wasting his time. He did some research into the

36. Hansen, *Arabia Felix*, p. 62.
37. 8 February, Forsskål, *Dagbok*, p. 5.
38. Petrus Bierchen, *Morbi expeditionis classicae*, Uppsala 1756.
39. 4 January–10 March, Forsskål, *Dagbok*, p. 3.

reasons for marine phosphorescence. He filtered the luminescent water through linen and paper, but he found no trace of organisms that could have produced the glowing of the sea. He also fished forms of sea-life out of the sea, amongst which he listed various species of rockweeds (*Fucus*) and sea lettuces (*Ulva*). These genera were, of course, recent Linnaean names, and Forsskål could not refrain from adding a note which dismissively explained that these species of Sjö-gräs (seaweeds) were merely known in Denmark with the collective name 'Tång'.[40] Notwithstanding these eager attempts at classifying whatever he managed to fish out of the sea, the ambitious young naturalist had to admit that he had not been able to find anything new so far. Still, these early activities in Danish waters confirmed that Michaelis' choice of a naturalist for the expedition had fallen on the right man. There are no records of Kramer's activities, and the only other member who is reported to have been working right from the beginning was Carsten Niebuhr, who was taking constant measurements with his astrolabe. The instrument had been especially calibrated by the self-taught astronomer Johann Tobias Mayer. Mayer had discovered the libration (wobbling) of the moon, and this had led in 1751 to his appointment as professor of economics and mathematics at the University of Göttingen where, as superintendent of the university observatory, he worked on lunar and solar tables, making it possible to determine longitude with precision. In Niebuhr he found someone who could furnish him with measurements with which to test his new methods of calculation. To kill time Niebuhr calculated that in two months the *Grönland* had covered 450 German miles (3,390 kilometres) but had only advanced about 5 miles (37.5 kilometres) in its voyage towards the Mediterranean.[41]

On 10 March it seemed that the expedition was finally on its way. The wind was good enough to cross the Kattegat and within two days the *Grönland* reached the North Sea. Once there, however, storms drove the ship off-course towards the Shetland Islands, and on 25 March they found themselves fast approaching Iceland. It would take another week before the weather improved and the ship could sail past Ireland towards Spain. During further gales another sailor fell overboard, and when they approached Cape Finisterre on 13 April, some of the men had died of scurvy. But although the voyage from Helsingör to the Mediterranean was far from comfortable, the members of the expedition did not lose heart or waste their time. Niebuhr was constantly measuring their geographic positions, while Forsskål was fully occupied in identifying molluscs, seaweeds, fish, birds and Crustacea, so much so that Niebuhr was full of admiration for his unflagging activity. To Linnaeus Forsskål wrote that he believed he had found numerous new medusae and two new genera of molluscs, which he asked Baurenfeind to draw.

On 13 May the *Grönland* anchored off the Provençal coast, a few miles from Marseilles. While waiting for the convoy to prepare for the further voyage to Istanbul, Forsskål explored parts of the coast around Marseilles, reported on the different soil types, and looked for plants between Castelnau and Gramont. Although these habitats had been visited and described by many other botanists, they were new to Forsskål. Moreover, while he recognized many European species,

40. Forsskål, *Dagbok*, p. 10.

41. Carsten Niebuhr, *Reisebeschreibung nach Arabien und andern umlegenden Ländern*, 3 vols., Copenhagen 1774, 1778, 1837, I.56.

he also had some foretaste of the kind of flora he imagined he would find in Asian and African countries. Being so near to Montpellier, Forsskål also decided to pay a visit to François de la Croix de Sauvages (1706–67), the first botanist at the University there. Sauvages was the author of *Methodus Foliorum* (1751), a system of classification of species based on the shape of leaves. The book was dedicated to Linnaeus. In his diary Forsskål wrote that Sauvages received him with a great display of affection ('ömhet') which, he added, was becoming to a person of his importance and in his diplomatic capacity in the service of the King of Denmark.[42] Forsskål did not, of course, give such explanations of Sauvages' warm welcome in his letters to Linnaeus. The truth was that Forsskål, who had not published anything significant in the field so far, knew that his importance depended mainly on his being one of Linnaeus' disciples. With Sauvages and his collaborator Antoine de Gouan he talked about Linnaeus' work, and he reported back to the master himself that Gouan was the only man at Montpellier who used the new system and who realized its true value, while Sauvages lamented that for three years he had not received a copy of Linnaeus' books.[43] Forsskål admitted to Linnaeus that he studied his books daily.[44] His complete devotion to his master is further illustrated by his perverse proposal to convey his important finds to him in secret. Forsskål had written to Bernstorff from Marseilles to ask if he could send seeds to important centres, including Uppsala. As all mail had to be sent via Copenhagen, he planned to refer henceforth to the numbering of genera and species in Linnaeus' work. In order to conceal his discoveries from the bunglers ('bönåsar') in Denmark, however, he warned Linnaeus that he would invert the first two digits of any number.[45] He would thus write 958.62 for *Cistus heliandethemum* instead of the correct 598.26.[46] Forsskål was much taken by this scheme of encoding, and he referred to it again from Istanbul more than two months later. But he justified all this in the name of true science, which Linnaeus loved so much, and which should not be allowed to suffer because there were people in Denmark suffering from unreasonable nationalistic pride.[47] Forsskål's subterfuge was only put to limited use, presumably because he was able to write to Linnaeus a month later that the Danish government had accepted his request to send seeds directly to the most important botanical gardens in Europe, including Uppsala.[48]

On 3 June the *Grönland* left Marseilles to escort the merchant vessels towards the Turkish coast. But not before Von Haven, who had travelled to the Mediterranean via Paris, where he took in a performance of Voltaire's *Tancrède*, had joined the other members of the expedition. The reunion was not a happy one. Von Haven, who had no special sympathy for his travelling companions, quarrelled violently with Forsskål. During a dinner on board the *Grönland* he insulted the Swedish politician Tessin whom Forsskål held to be a decent man. Forsskål said so, and added that he who abused Tessin's reputation must be a man of low reputation himself. To this

42. Forsskål, *Dagbok*, p. 18.
43. Forsskål to Linnaeus, 27/5/1761, in Fries, Hulth and Uggla, *Bref och skrifvelser,* p. 141.
44. Forsskål to Linnaeus, 1/8/1761, in Fries, Hulth and Uggla, *Bref och skrifvelser,* p. 142.
45. Forsskål to Linnaeus, 27/5/1761, in Fries, Hulth and Uggla, *Bref och skrifvelser*, p. 140.
46. Forsskål to Linnaeus, 1/8/1761, in Fries, Hulth and Uggla, *Bref och skrifvelser*, p. 142.
47. Forsskål to Linnaeus, 1/8/1761, in Fries, Hulth and Uggla, *Bref och skrifvelser*, p. 143.
48. Forsskål to Linnaeus, 4/9/1761, in Fries, Hulth and Uggla, *Bref och skrifvelser*, p. 146.

Von Haven is reported to have hurled some obscenities at Forsskål.[49] This incident was to create much mischief and delay later on.

On 14 June the convoy reached Malta, where it anchored for six days. On the island, Forsskål busied himself with the study of local natural history. Although Niebuhr says that he and Forsskål made an excursion to the Salines, his description of Malta is mainly restricted to comments on the oriental style of building and the splendour of its churches.[50] Forsskål's diary does not supply much more natural historical information. Like Niebuhr he indulges in describing in detail the cave in which St Paul was supposed to have hidden during the months following his shipwreck on the island, and he too seems fascinated by the alleged miraculous power of the rocks inside. Ashamed of this digression, he closes his entry for Malta by saying that the island's natural history was much more important to him than the debate between the superstitious and the sceptics about the veracity of Paul's lot on Malta: 'I collected a great many plants on its rocks,' he concludes.[51] He refers in brackets to his 'Flora Melit.' He also mentions a list of fish in the Maltese waters which he got from a learned but unnamed friend,[52] who is described as a prominent physician in the *Catalogus Piscium Melitensium*, and whom the historian Georg Zammit Maempel surmises to have been Giorgio Giovan Battista Locano.[53]

Malta was no more than a promising start for Forsskål's research on fauna and flora. Maempel assesses the importance of his scientific work for Malta as that of 'one of the founders of local palaeontology and of the study of other fields of natural history of the Maltese Islands'.[54] Not too much importance should be read into this phrase. Admittedly, Forsskål's six days collecting specimens resulted in the first 'scientific' list of invertebrate fossils on the island. Yet, this list contained only 11 molluscs and 3 echinoids. Again, his 'Florula Melitensis' occupies a mere two pages in his larger *Flora Aegyptiaco-Arabica* and he only recorded 87 species of plants, 9 of which were cultivated. Maempel's claim that seven plants in Haslam and Wolseley's *Flora of the Maltese Islands* (1977) still carry Forsskål's name is misleading too.[55] None of these plants were actually described by Forsskål from plants on Malta. With one possible exception,[56] his descriptions of these species were based on specimens found later in Egypt and the Yemen.[57]

49. See Hansen, *Arabia Felix*, p. 73.
50. Niebuhr, *Reisebeschreibung*, I.61.
51. 'Jag samlade ihop en stor mängd örter på dessa klippor', Forsskål, *Dagbok*, p.27.
52. 'En lärd vän', Forsskål, *Dagbok*, p. 27.
53. George Zammit Maempel, 'The Arabian Voyage of 1761–67 and Malta: Forsskål and his Contribution to the Study of Local Natural History', *Proceedings of History Week*, 1992, pp. 60–4.
54. Maempel, 'The Arabian Voyage of 1761–67 and Malta', p. 74.
55. Maempel, 'The Arabian Voyage of 1761–67 and Malta', p. 57.
56. His *Cnicus dentatus* (modern name: *Carthamus dentatus* (Forssk.) Vahl) is not included in the 'Florula melitensis'. Although we cannot exclude the possibility that his description (Forsskål, *Flora Aegyptiaco-Arabica*, p. 217) was based on specimens from Malta, all it says about the island is: '*Maltae* frequens in ruderatis', while it adds 'In *Archipelago* rarior'.
57. *Suaeda vera* was found by Forsskål on Malta, but he failed to publish it. Forsskål's description is based on a specimen from Alexandria.

Istanbul

On 20 June the *Grönland* left the harbour of what is now Valetta, sailed just south of the Peloponnese, crossed the Cyclades, and headed north-east for Izmir, where the ship dropped anchor on 3 July. Forsskål exulted: 'We arrived so early that on the very same day I could go and look for plants for the first time on Asian soil.'[58] The following week Forsskål assiduously collected plants and shells. He noted that molluscs were particularly abundant on the Izmir coast.[59] In his diary he also reports on what he thought was a new species of bird and to which he gave the name *Turdus seleucus*. The local people told him that it was famed for killing large quantities of locusts and that the Koran consequently forbade the killing of it.[60] He devotes some space to the edible molluscs that could be found around Izmir and how they should be cooked. The juxtaposition of such disparate pieces of information reminds the reader that the Danish expedition also aimed at gathering scientific facts which could serve potential commercial interests.

When, on 13 July, the *Grönland* reached the island of Tenedos, the members of the expedition were asked to leave the warship, which would not proceed any further because the plague had broken out in Istanbul. They were picked up by a Turkish vessel, which only made its way to the Ottoman capital very slowly—so slowly that Forsskål had many opportunities to collect specimens of plants and shells on land. He added to his botanical lists some 40 species from Tenedos, 73 from Imros, and 111 from the area around the Dardanelles. The gardens at Istanbul provided him with further material, as did an excursion to Rhodes.[61] This enabled him to draw up a 'Flora Constantinopolitana, littoris ad Dardanellos et insularum Tenedos, Imbros, Rhodi' in which he included 481 species. Whereas Forsskål's 'Florula littoris Gallie' and his 'Florula Melitensis' were, with a few exceptions, mainly lists of species, his flora of the Turkish coast was based on more extensive exploration. Many of his entries include technical descriptions, while special attention is paid to habitat, soil, and geographical position. Forsskål also records numerous local Greek plant names, which, Niebuhr explains in the preface to *Flora Aegyptiaco-Arabica*, he heard from peasants as well as from a Greek who had a good knowledge of pharmacology.[62] The preface also draws attention to the fact that he found new species among the many plants he studied along the Turkish coast. Indeed, in the 'Flora Constantinopolitana' 45 names are marked as new species. This number indicates the spirit in which he carried out his research: almost one out of every ten plants he found was new to him. It convinced the emerging botanist that he was indeed performing groundbreaking work, and he had high hopes for what Egypt and the Yemen had in store for him.

Very few of Forsskål's new names in the 'Flora Constantinopolitana' can be accepted today, however. Almost a third describe obscure and unidentified species,

58. 'Vi anlände i så god tid, at jag kunde än sama dag för första gång botanicera på Asiatisk botn.' Forsskål, *Dagbok*, p. 27.

59. 'Orten är myket öfwerflödande på allehanda arter af Snäckor', Forsskål, *Dagbok* p. 28.

60. Forsskål, *Dagbok*, pp. 28–9; Hansen has pointed out that the bird is not mentioned in the Koran (*Arabia Felix*, p. 77); today the bird is known as *Pastor roseus*.

61. Rhodes was visited during a stop on their way from Istanbul to Alexandria.

62. 'Nomina Graeca saepe adjecta, incertae auscultationis non sunt. Partem eorum a rusticis habuit: partem aliam a Viro Graeco Rei Pharmaceuticae perito.' Forsskål, *Flora Aegyptiaco-Arabica*, p. 16.

and in the absence of preserved specimens the International Botanical Congress of 1993 recommended that the names should not be adopted if they conflict with widely recognized ones.[63] Approximately 10 per cent of the new species still bear Forsskål's name. These comprise: *Taraxacum megalorrhizon*,[64] *Galium paschale*,[65] *Bupleurum flavum*,[66] *Polystichum setiferum*,[67] *Ephedra foeminea*.[68] Most of the other plants had already been published by Linnaeus under different names in 1753. A few species were published by Linnaeus in 1767,[69] i.e. after Forsskål had found and described them, but before Niebuhr could officially publish them in 1775. Forsskål's *Salvia bifida*[70] is one of these: Linnaeus's description in 1767 was based on specimens he had grown in Uppsala from seed Forsskål had sent him. He generously decided to call it *Salvia forskahlei*.

On board the Turkish vessel to Istanbul the Danish scholars were to experience 'how one travelled with Muslims'. Before leaving the *Grönland* they had had some foretaste in a series of encounters with Turks who came to taste Fischer's wine. Forsskål remained silent about them, but Niebuhr commented that their language, clothes and behaviour were so strange to them that they had little hope of finding in the Levant much pleasure or friendliness. It might have been the awareness of being so different that contributed to the decision to wear oriental dress and to give up the use of chairs and other European comforts.[71] The long stay at Istanbul till 8 September, however, was not amongst Turks or in Turkish style, as they were invited to stay at the large house of the Royal Danish Envoy Gähler, where Niebuhr, who had been suffering from dysentery since leaving Malta, could fully recover. Gähler procured for them a travel permit from the Turkish Sultan, as well as a series of letters of introduction for Egypt. He also handed the members of the expedition more detailed instructions for research in Egypt and the Yemen.

Michaelis' 'Fragen'

Our present knowledge today of the questions that were to lead the scholars in their research is based on their formulation by Michaelis when he prepared them for publication in 1762 as *Fragen an eine Gesellschaft gelehrter Männer die auf Befehl Ihro Majestät des Königes von Dänemark nach Arabian reisen*. This work became a kind of handbook for biblical questions on the Arabian peninsula. Soon after it appeared it was translated into French and Dutch.

The *Fragen* contains 100 numbered questions which range from simple queries

63. The XV International Botanical Congress urges 'plant taxonomists [...] to avoid displacing well established names for purely nomenclatural reasons, whether by change in their application or by resurrection of long-forgotten names' (Preface to the *International Code of Botanical Nomenclature (Tokyo Code)*.

64. Originally described by Forsskål as *Leontodon megalorrhizon* (Forsskål, *Flora Aegyptiaco-Arabica*, p. 216).

65. Forsskål, *Flora Aegyptiaco-Arabica*, p. 203.

66. Forsskål, *Flora Aegyptiaco-Arabica*, p. 205.

67. Originally described by Forsskål as *Polypodium setiferum* (Forsskål, *Flora Aegyptiaco-Arabica*, p. 185).

68. Forsskål, *Flora Aegyptiaco-Arabica*, p. 219.

69. Carolus Linnaeus, *Mantissa plantarum*, in the appendix to the twelfth edition of *Systema naturae*, Stockholm, 1767.

70. Forsskål, *Flora Aegyptiaco-Arabica*, p. 202.

71. Niebuhr, *Reisebeschreibung nach Arabien*, I.78.

Title page of Johann David Michaelis, *Fragen an eine Gesellschaft gelehrter Männer, die auf Befehl Ihro Majestät des Königes von Dänemark nach Arabien reisen*, Frankfurt 1762
The Arcadian Library, 12383

Johann David Michaelis,

Königl. Großbritannischen und Churfürstl. Braunschweigisch-Lüneburgischen Hofraths, Professors der Philosophie zu Göttingen, und Directors der Societät der Wissenschaften daselbst,

Fragen

an eine Gesellschaft

Gelehrter Männer,

die auf Befehl

Ihro Majestät des Königes

von Dännemark

nach Arabien reisen.

Frankfurt am Mayn,

bey Johann Gottlieb Garbe,

1762.

about place, plant or animal names to complex and fully articulated disquisitions on diseases in the Near East. The expedition was also asked to investigate various geographical and meteorological matters, to observe practices in agriculture and commerce, and to shed light on some Western perplexities about the sexual habits of Muslims. Michaelis thus asked whether the Jews in Exodus could have crossed the Red Sea by a narrow, but now submerged (coral) isthmus. He sought information concerning machinery for irrigation and inquired into the construction of huts. He asked detailed questions about the manifestations of such diseases as leprosy, scabies of the face, and elephantiasis, and whether houses and clothes could be sources of contamination of leprosy. He showed interest in the oriental methods of castration and circumcision, as well as their habits of preserving virginity and having sexual intercourse during the period of menstruation. He asked about clean and unclean animals. He wanted to know whether flying fish in the Red Sea flew against the wind, and how the Arabs counted the feet of locusts. He wondered about the existence of other flying animals. Did flying quadrupeds exist? And flying snakes? And what about the two-legged mountain mouse? Although there were plenty of questions for each member of the expedition, a considerable number are about animals and plants, and these would naturally fall to Forsskål's competence.

Michaelis states at the outset of his *Fragen* that nearly all his questions served in the first place for the elucidation of Holy Writ. 'I can foresee that this [approach] will displease some,' he apologizes, '[but] they should not pick on me, when I admit that everything seems important to me that can clarify the true meaning of a book on which all our religion is based.' The study of the Old Testament, he emphasizes, 'forces us to investigate fully all the natural history and customs of the Levant':

It contains nearly three hundred names from the vegetable kingdom, countless ones from the animal kingdom, and quite a few names of gems: it is entirely woven through with customs and geography of the Levant.[72]

Michaelis' book is in fact a remarkable compendium of mid-eighteenth-century knowledge of the Arab world. Its scope went far beyond linguistic or philological investigation. The *Fragen* does not merely reveal the lacunae in western knowledge: in their elaborate formulation, the questions also reveal the extent of European knowledge of Arab customs, geography and natural history. This was no small feat in itself, and indeed Michaelis admitted that, even if the questions were no more than a record of his own ignorance, they had still cost him more effort than any other book he had ever published.[73]

Although Michaelis had initially felt disappointed with the number of scientific questions he received following his request in the newspapers, further letters had reached him during the seven months which had elapsed since the *Grönland* had left Denmark. If Michaelis had initially wanted the advice of international scholars on what to include in the questions, he had by now made so much progress on the draft

72. 'Beynahe dreyhundert Namen aus dem Gewächsreiche, ich weiß nicht wie viele aus dem Thierreiche, and ziemlich viele Namen der Edelgesteine, kommen in demselben vor: mit Sitten der Morgenländer, mit Geographie ist es ganz durchflochten.' Michaelis, *Fragen*, 'Vorrede', [pp. 28–9].
73. Michaelis, *Fragen*, 'Vorrede', [p. 36].

Johann David Michaelis,
Frage I: What is *suph*?
Fragen an eine Gesellschaft gelehrter Männer, die auf Befehl Ihro Majestät des Königes von Dänemark nach Arabien reisen,
Frankfurt 1762, p. 1
The Arcadian Library, 12383

NEXT OPENING
Johann David Michaelis, Frage LXXI: With what disease did God threaten the Israelites in Deuteronomy 28:27?
Fragen an eine Gesellschaft gelehrter Männer, die auf Befehl Ihro Majestät des Königes von Dänemark nach Arabien reisen,
Frankfurt 1762, pp. 196–7
The Arcadian Library, 12383

of his *Fragen* that he did not need, and probably did not want, to include this extra material. The book was going to be *his* alone. The letters were therefore forwarded to Istanbul, where Gähler handed them over to the members of the expedition.[74] He also gave them the first draft of Michaelis' questions.

The final version of the *Fragen* as we have it never reached the expedition. Niebuhr explains in the 'Advertisement' to his *Beschreibung von Arabien* (1772) that, apart from a document drafted by the French Académie Royale des Inscriptions et Belles Lettres, the expedition had left with no more than two short questions from Michaelis, and that the hundred questions which were later published in the *Fragen* only reached them by fits and starts, the first in Istanbul, some more in Egypt, and the rest in the Yemen. The printed version only reached Niebuhr in 1764 in Bombay, more than a year after the deaths of Von Haven and Forsskål, and some time after the deaths of Baurenfeind and Kramer.[75] The members of the expedition probably saw most of the questions in the end, but not in their final form. For at least two years Michaelis worked assiduously on the draft of his book, inviting to his house in Göttingen learned friends to whom he could read the questions out in the evenings. One thing the friends objected to was that Michaelis asked far too many questions for five travellers to answer in three years. Although the published version features only 100 numbered questions, each of them contained further related ones, so that the ultimate number was much nearer a thousand. Michaelis protested that that was part of his design: it was most unlikely that all the questions would find answers, he admitted, but the more he asked, the more he could hope to get answered, especially as, on such an expedition, finding answers to at least half the questions would mainly depend on luck and circumstance.

In the *Beschreibung von Arabien* four pages of preliminaries are printed with Forsskål's answers to Michaelis' 100 questions, which, Niebuhr thinks, will please scholars. They were put together from the notes on loose sheets found by Niebuhr after his return to Europe, and which he ordered according to the final numbering of the 1762 version. Two things strike the reader who compares them with Michaelis' voluminous treatise: their brevity and the many medical questions that would naturally have been addressed to Kramer.

Most of the answers Forsskål gives are short remarks, as for example the answer to the first question: in the three pages of text accompanying the question about *suph*—'What is סוף, from which the Red Sea derives its name? and of that which grows in the Nile?'—Michaelis asks for a detailed botanical description, as, he maintains, this cannot be found in the works of Celsius, Robert Southwell, Cristóvão da Costa or Pliny. He also asks whether it differs from the Sargasso Weed, whose berries, one of Michaelis' friends had pointed out, gave stretches of sea a red colour. Forsskål does not provide any description in Niebuhr's list, nor does he answer any of the more specific questions. He limits himself to a mere: '*Fr.* I. סוף Planta aquatica, ar. دیس'.[76]

Another example illustrates the striking contrast between Michaelis' elaborate and

74. 'Sie haben mit den Fragen, welche ich in diesem Buch aufwerfe nichts zu thun, sondern sind den Reisenden in Orginal mitgegeben.' (Michaelis, *Fragen*, 'Vorrede', [p. 17]).

75. Carsten Niebuhr, *Beschreibung von Arabien: Aus eigenen Beobachtungen und im Lande selbst gesammleten Nachrichten abgefasset*, 2 vols., Copenhagen 1772, I.xvi; Cf. Michaelis, *Fragen*, 'Vorrede', [p. 30].

76. Michaelis, *Fragen* pp. 1–4; Niebuhr, *Beschreibung von Arabien*, I.xliii.

I.

Was ist das סוף, davon das rothe Meer den Nahmen hat? wie auch dasjenige, so im Nil wächset?

Von diesem, meiner Meinung nach gedoppelten Suph, wird eine genaue botanische Beschreibung und Abzeichnung, wie auch eine Nachricht von der Farbe des im rothen Meer befindlichen Suph verlanget, weil man sich in folgendes nicht finden kann.

Celsius hat in seinem hierobotanico P. II. p. 65. seqq. wahrscheinlich genug gemacht, daß das Suph eine alga oder Meergras sey. Ein geschickter Jesuite, der selbst in Aethiopien und auf dem rothen Meer gewesen war, und welchen Robert Southwell zu Lißabon gesprochen hat, erzählte diesem noch etwas näheres, so Celsio unbekannt geblieben. *) Er hatte auf

Suph ist ein Alga

*) Recueil de divers Voyages faits en Afrique et en Amerique. (à Paris 1674) pag. 232.

Mir ist beygefallen, ob es nicht vielmehr der Stickfluß sey, der von خرّ schnarchen genannt seyn könnte. Zwey im Orient zu entscheidende Fragen werden dieß ausmachen können:

1) Wird خرّ auch von dem Röcheln eines Stickenden gebraucht, oder gehet es blos auf das Schnarchen der Schlafenden.

2) Saadias übersetzt unser Wort فَالِج. Brauchen die Araber dieß Wort, so Golius Paralysis und Hemiplexia übersetzt, von der Paralysi der Lunge, die einen Stickfluß zu wege bringt?

Der Samaritaner übersetzt דקיקתא. Was ist dieß für eine Krankheit? Ich frage hier immer, was es nach dem Gebrauch des Wortes für eine Krankheit ist, nicht aber was für eine es der Derivation nach bedeuten könnte.

LXXI.

עפלים Deut. XXVIII, 27. 1 Sam. V, 5. 6. 9. 12. VI, 5. 11.

Gott drohet den Israeliten eine Krankheit, die entweder עפלים heißt, oder den also benannten Theil des Leibes betreffen soll,

5. B.

5. B. Mos. XXVIII, 27. und dieß Uebel fühlen die Philister 1 Sam. V. und VI. Ob עפלים die Krankheit selbst, oder ihren Sitz bedeute, ist völlig ungewiß. Bochart hat von dieser Plage P. I. B. II. Cap. 36. gehandelt und entscheiden wollen, allein wenn er auch so glücklich seyn sollte Wahrheit zu haben, so überzeuget er doch nicht. Die Masorethen haben am Rande davor טחרים, und dieß kommt auch 1 Sam. VI, 11. im Text vor. Bochart unterscheidet es von עפלים, und meynt עפלים sey die Krankheit, טחרים aber der Sitz der Krankheit: und doch scheint aus Vergleichung des fünften Verses eben des Capitels klar zu seyn, daß beyde Wörter völlig einerley bedeuten. Im Syrischen und Chaldäischen ist טחור gebräuchlich gewesen, wie man aus den Uebersetzungen siehet: und dieß kann freylich 1 Sam. VI, 11. in der syrischen Uebersetzung nichts anders als der kranke Theil des Leibes seyn, nicht aber die Krankheit selbst.

Da die aus Büchern geschöpfte Kenntniß des Arabischen nicht so weit gehet, diese Namen zu erklären, so bitte ich die Reisenden zuförderst darauf zu sehen, ob die geredete arabische Sprache nichts zu Aufklärung der beyden Wörter טחור und עפול, in so ferne sie Krankheiten oder ein Glied des Leibes bedeuten, darreichet.

Fr. XLIII. אהלים Num. 24, 6 مضارب Saadias i. e. prosapia genus. Karaitis in Kahira, Sandal.

Fr. XLVI. قحم רְאֵם Káhm. Muri. بردي גמא Saadias. Exod. 2, 3.

Fr. XLVIII. شبح in montibus Jemenis occurrit. Muri.

Fr. XLIX. לענה Karaitis in Káhira علقم alkam.

Fr. LIV. פצוע דכה sine testibus (Karaiti in Káhira). כרות שָׁפְכָה sine virga & testibus. כתות zerdrückt. Castrati demtis & testibus & virga dicuntur مخصي

Fr. LVIII. De sputo ob negatum Leviratum, 5. Mos. 25, 9. Etiamnum mos servatur a Judæis Jemenis. Sie binden einen Schuh an den Waden. Diesen löset die יבמת mit der rechten Hand ab, wirft ihn auf die Erde, und speiet dreymal vor ihm auf den Boden, nicht in sein Angesicht. Sie erklären בפניו durch לפניו Nec patres in filiorum vultum spuunt in Jemen, nec de lepra phrasis illis est: Deus in vultum ejus exspuit.

Fr. LIX. Bey der translatione dominii den Schuh auszuziehen, ist weder bey den Juden noch andern Arabern länger gebräuchlich.

Fr. LXI. Phrasis Saadiæ جد عقب non in usu est arabice loquentibus. Muri.

Fr. LXII. שפיפון Karaitis in Káhira arab. مقرون Ita & Saadias.

Fr. LXIV. פקעות 2. Reg. 4, 39. حنظال Colocynthis. Muri.

Fr. LXVII. שחפת سل Sill. Saadias.

Fr. LXVIII. קדחת حرارة Saadias.

Fr. LXXI. Dolor dentium frequens in Jemen, dicitur وجع الضرس עפלים بواسير Hæmorrhoides. Muri.

detailed questions and Forsskål's more than concise answers. In Frage 71 Michaelis asks about the disease with which God threatened the Israelites in Deuteronomy 28:27. It is entirely unclear, Michaelis writes, whether the word describes the disease or the afflicted part of the body. He compares it to the word which the Masoretes annotated in the margin of the text, a word which is also used in 1 Samuel 6:11. For Michaelis the biblical scholar and oriental linguist Samuel Bochart is not convincing in his distinction between words indicating the disease and the afflicted part of the body. In the Syriac version, Michaelis adds, there is a related word, which can only refer to the diseased part of the body. 'As the knowledge of Arabic drawn from books', Michaelis writes, 'does not go far enough to explain these nouns, I entreat the travellers to observe attentively if spoken Arabic can offer some clarification as to the meaning of both the words [... and see] whether they refer to the disease or the afflicted part of the body.' He insists, moreover, on finding out whether the disease which afflicts the part of the body these words refer to is to be identified as: (1) a prolapsus of the intestine; (2) dysentery; (3) haemorrhoids; (4) phagedaena. With respect to these diseases Michaelis also asks whether the use of the Arabic 'shatara' can shed any light on the question, and why some exegetes should have used other Arabic words to explain the disease.[77] Here too the complexity of Michaelis' argument seems to have been evaded by the concise and uninformative nature of Forsskål's answer in Niebuhr's *Beschreibung*.[78] Niebuhr clearly reproduced such notes as Forsskål had jotted down in the field during their expedition, or in conversations with local Arabs, and although Michaelis' original questions obviously did not contain all the additional queries we have today in the *Fragen* of 1762, one cannot avoid the impression that Forsskål thought the questions a waste of time and took little trouble to answer them.

Sample page of Forsskål's answers to Michaelis' questions. Carsten Niebuhr, *Beschreibung von Arabien: Aus eigenen Beobachtungen und im Lande selbst gesammleten Nachrichten abgefasset*, Copenhagen 1772, p. xlv
The Arcadian Library, 10525

Question 71 is characteristic of most of the *Fragen*. Although Michaelis' queries often touch on matters of Biblical exegesis, they do not in fact go beyond the linguistic problems. The interest in the linguistic aspect of a Biblical passage is often at the expense of its general meaning. In a footnote at the end of his four-page treatise on the linguistic mysteries of the disease Michaelis adds: 'While reading this out to my friends, Mr D. Heilmann reminded me of the fact that it is ununderstandable how haemorrhoids could have been held as proof of an offended divinity.'[79] It creates a vivid image of the evenings at Michaelis' house in Göttingen when the group of (medical) friends were so absorbed in solving the linguistic problems of the Old Testament that they at times forgot the broader context of what they were really talking about.

Niebuhr's list of Forsskål's answers to Michaelis' 100 questions remains unimpressive. Most of the *Fragen* are hardly dealt with. Forsskål provided 33 answers; 13 of these were based on the oral information he got from a Jew by the name of Muri at Mocha. It is clear that Forsskål's achievements as a naturalist are not based on the answers to the printed questions. His scientific reputation rests on the

77. Michaelis, *Fragen*, pp. 196–200.
78. Forsskål's full answer is as follows: '*Fr.* LXXI. Dolor dentium frequens in Jemen, dicitur وجع الضرس ץפלים بواسير Haemorrhoides. Muri.' ('Toothache in the Yemen is often called wajaʿ al-dirṣ; tsefilim bawāsir are haemorrhoids. Muri.' Niebuhr, *Beschreibung*, I.xlv.
79. Michaelis, *Fragen*, p. 199.

descriptions of the numerous hitherto unknown species, and his vast collection of vegetable and animal specimens, much of which reached Copenhagen safely after his premature death in the Yemen.

Alexandria and Rosetta

Apart from some secret communication with young slave-girls in a cabin above, the sea voyage on board a Turkish vessel from Istanbul to Alexandria was uneventful. But all was not well. Forsskål, Niebuhr and Baurenfeind were worried. Relations between Forsskål and Von Haven had not improved since their quarrel in Marseilles. In Istanbul Gähler had tried to reconcile the two scholars, but in vain. While at sea Kramer could no longer keep to himself the fact that he had seen Von Haven swear revenge on Forsskål soon after the forced reconciliation at the envoy's house. Kramer, moreover, had noticed that among the medicines Von Haven had ordered from a chemist in Istanbul there was a considerable quantity of yellow and white arsenic. To the other men there was no doubt that this meant that Von Haven had decided to poison Forsskål.

On 21 September the ship dropped anchor in the harbour of Rhodes, and Forsskål, Niebuhr and Baurenfeind made for the house of the French Consul, where they were at first refused admission because of their Turkish dress. It was here that they met a gentleman who was willing to carry a letter to Gähler, in which they revealed Von Haven's sinister plans. They asked for royal orders to have the philologist removed from the expedition. The letter was written and signed by Forsskål and co-signed by Niebuhr and Baurenfeind.

There was hardly time to do any serious scientific work during the one day in Rhodes, but with the Consul's interpreter as a guide they managed to make some little excursions on the island. Forsskål collected a dozen plants, but he had already seen most of the species he found on Tenedos and near Istanbul.[80] They were included in his 'Flora Constantinopolitana'. He glosses over the brief visit to Rhodes with 'in transitu, paucis horulis visitavimus' ('in transit, we only spent a few short hours there'), even if his list of plants included one new species, which he called *Anthericum aphyllum* and of which he furnished a valid description.[81] The plant, however, has not been identified with any known species today.

Early in the morning of 22 September the ship started on its last lap of sea that divided them from Egypt, and they reached Alexandria in the evening of 26 September. From a distance the city's palaces and mosques were a fine sight ('ein schönes Ansehen'), but the members of the expedition felt gloomy. Apart from Von Haven's menace, the prospect that lay before them was upset by other considerations. During the brief crossing some of the Turkish sailors fell ill and died, presumably of the plague. The event impressed upon the members of the expedition the fact that travelling in the Near East imperilled their lives directly. Kramer visited the dying sailors, but the disease was not transmitted to the Europeans, something which made

80. Forsskål, *Dagbok*, p. 42.
81. Forsskål, *Flora Aegyptiaco-Arabica*, p. 209.

the stoic Niebuhr doubt whether the cause of death was indeed the plague.[82] That the danger of disease was real enough would be proved by the disastrous events that unfolded a year and a half later in the Yemen.

In Alexandria the members of the expedition were also confronted with the question of personal safety during their scientific excursions. Soon after their arrival they witnessed a ferocious scuffle between a band of wandering Arabs and the inhabitants of the city, a fight which started with a volley of stones and ended in gunfire. If these Arabs swarming around the city could cause such damage, the Europeans realized, travelling about the country could expose them to considerable danger. Niebuhr records that when, on a later occasion, Forsskål went looking for plants between Cairo and Alexandria, he was stopped by a band of robbers. He had to hand over all his possessions. 'It must be considered a true kindness in them', Niebuhr drily adds, 'that they gave him his trousers back.'[83]

But, in spite of the dangers posed by Von Haven when he was in his company and those from the bands of Bedouin while studying the plants in the countryside, Forsskål could not but feel exhilarated when he saw the Egyptian flora. In his diary he described the feelings of a botanist who has experienced little other than northern climes and who had reached Egypt after a long and monotonous sea voyage. The novelty of it all—the people, the places, the nature, the plants—created sensations which could not be put into words. 'I could think of nothing other than collecting and observing!'[84] He found that the safest place to collect specimens on his own was in what he decided to call the 'botanic O'—by which he meant the semicircular area between Alexandria and Rosetta—as the Bedouin did not venture into the settlements on the fringe of the isthmus. Although limited in his movements, Forsskål was more than satisfied with the variety of habitats in his 'botanic O', with its beaches, hills and valleys, streams, farmlands, shaded groves and open fields.

In its scientific pursuits the Danish expedition had also to reckon with the local population, whose curiosity and general inability to understand the men's intentions often severely hindered their research. But this was taken in good spirit. Niebuhr, for example, was amused by the general agitation he caused when making geographical observations with his astrolabe. When he let the natives have a look through the instrument and they saw the world turned up-side-down, they felt alarmed. A rumour soon spread that he had come to Alexandria to overturn the city. It is a measure of Niebuhr's unbiased way of reporting that he concluded this passage with: 'One should not marvel about the fact that the Muslims were diffident about what they saw [in the astrolabe], as only recently many people in Europe considered anything they failed to understand as magic.'[85]

82. Niebuhr, *Reisebeschreibung*, I.89.

83. Niebuhr, *Reisebeschreibung*, I.102.

84. 'Jag hant ej mer än samla och betrakta!' Forsskål, *Dagbok*, p. 64.

85. 'Man darf sich eben nicht sehr verwundern, daß die Mohammedaner über dergleichen Beobachtungen argwöhnisch werden, da man nicht vor langer Zeit auch noch Europäer genug gefunden hat, die alles für Zauberei hielten, was sie nicht gleich begreifen konnten.' Niebuhr, *Reisebeschreibung*, I.96. Robert Heron clearly did not share this sympathy for the Arabs. In the English adaptation of Niebuhr's work, published in 1792, he did not hesitate to render his conclusion as follows: 'The [...] rudeness of these Arabians were not the only circumstances that obliged me to repress my curiosity. The stupidity and ignorance of the native inhabitants, who viewed my measuring apparatus with distrust and fear, were not less unfavourable to the success of my enquiries and observations.' Robert Heron, *Travels through Arabia, and Other Countries in the*

Georg Wilhelm Baurenfeind, illustration of three *Salsola* species in Linnaeus' personal copy of Forsskål's *Icones* (plant B is *S. inermis*).
Icones rerum naturalium: quas in itinere Orientali depingi curavit Petrus Forskål. Edidit Carsten Niebuhr, Copenhagen 1776, plate VIII
Linnean Society Library, BL.690

Linnaeus had urged Forsskål to pay particular attention to the mythical Meccan Balsam Tree and see if he could determine what genus it belonged to. He made it clear that it was his special wish to see this done before he died. Although there are numerous references to *Opobalsamum* in antiquity, the tree that produced it was extremely rare, and as a consequence its precise identity had remained uncertain. It is mentioned in the Book of Genesis as one of the precious goods the Ismaelite trading caravans carried into Egypt (37:25), and it is among the gifts Jacob's sons took to Joseph (43:11). Its healing properties are pointed out in various passages in Jeremiah (8:22; 46:11; 51:8), while in Ezekiel it is included among the riches and commerce of Judea (27:17). Pliny the Elder, in his *Naturalis historia*, wrote in a famous passage that

formerly it grew in only two gardens, both belonging to the king; one of them was not more than twenty iugera in extent and the other less. This variety of shrub was exhibited to the capital by the emperors Vespasian and Titus; and it is a remarkable fact that ever since the time of Pompey the Great even trees have figured among the captives in our triumphal processions.[86]

Diodorus Siculus wrote that the balsam tree could only be found in a small area near the Dead Sea and nowhere else in the world. Perhaps even better known than these references was the popular story to which Flavius Josephus referred,[87] namely that the Queen of Sheba brought the plant as a gift to King Solomon, and that it subsequently became one of the country's most precious products and an important source of royal revenue. Since the *Opobalsamum* had thus enjoyed much fame throughout history, it is not surprising that Linnaeus should have been eager to establish the identity of this rare and magic plant and place it in his new taxonomical system.

One of the first things that Forsskål enquired after in Alexandria was thus whether anybody could guide him to a specimen of this mysterious tree. In his diary he recorded how one day he was told of a tree in a garden outside the town, and that he lost no time in going to the garden indicated, only to find an empty space where an unidentifiable tree had once stood. On another day a Jew was mentioned to him who was supposed to have the rare plant in a pot. Again Forsskål rushed off, venturing into the dark alleys of the Jewish quarter of the town. The visit was again futile. It was not the coveted Opobalsamum Tree. All he found was a plant which in Arabic was called *ballesan*, and which he identified as *Mormodica balsamina* (Balsam Apple), a luscious member of the cucumber family whose pulpy aril around the seeds is eaten in the East or used for the treatment of rashes of the skin. The moment had not yet come, Forsskål concluded, when *Opobalsamum* would be seen and described,[88] and to Linnaeus he wrote: 'I dare say I will see the Balsam Tree yet in Cairo.'[89]

There was no occasion to feel disappointed. The excursions around Alexandria had yielded Forsskål a wealth of plants, which included a number of new species

East, Performed by M. Niebuhr, 2 vols., Edinburgh 1792, I.38–9.
86. Pliny, *Natural History*, transl. by H. Rackham, The Loeb Classical Library, 10 vols., London, Heineman 1968, XII.liv, pp. 79–81.
87. *Antiquities of the Jews*, viii.6.
88. Forsskål, *Dagbok*, pp. 63–4.
89. Forsskål to Linnaeus, 20/10/1761, in Fries, Hulth and Uggla, *Bref och skrifvelser*, p. 152.

C
B
A

of *Salicornia*, *Salsola*, *Senecio*, *Statice*, *Passerina*, *Zygophyllum*. Not all these carry Forsskål's name today, but *Salsola inermis*, *S. longifolia*, and *S. tetrandra* do. The first of these salt-loving succulents was drawn by Baurenfeind and reproduced in Plate VIII of Forsskål's *Icones rerum naturalium*.

The month of delay in Alexandria before continuing their journey to Cairo, and then to the Yemen, was presumably caused by the answer the men expected to their letter to Gähler about Von Haven's sinister intentions. But as no letter came, the members of the expedition decided on 31 October to move on to Cairo, and wait there for further royal instructions. Just as it was unsafe to look for plants outside the gardens of Alexandria, it was equally impossible to travel by land to Cairo without the protection of a caravan against the dangers of wandering Bedouin robbers, with whom the country seemed to abound, as Forsskål complained to Linnaeus.[90] Niebuhr, too, wrote in his travel account that they had all wanted to avoid the habitual itinerary up the Nile from Rosetta to Cairo, and go by land instead. However, it soon became clear to them that this was impossible because of the Bedouin. They thus boarded an Egyptian vessel to Rosetta, where Forsskål was amazed by the glorious scenery of the Nile Delta and the stunning views from a flat tower just outside Rosetta which, in its summer splendour, were 'unequalled among all the vistas that one can have around here'.[91] Baurenfeind drew one of his finest panoramas there.

On 6 November they continued towards Cairo. Even on the river they were not entirely free from robbers by night. These were not to be feared, however, as long as a guard made it clear that he was equipped with a gun, something which could be achieved by assiduously firing it.[92] Notwithstanding these restraints, Forsskål tried to describe what he could observe from the river and, because of the slow movement of the boat upstream, he managed to go on land every now and then and collect specimens during the four days' journey.

Cairo

In Cairo the members of the expedition found separate accommodation. Von Haven stayed with an influential French merchant, Kramer accepted the invitation of the Dutch consul and stayed at his house, Baurenfeind found a room with the Capuchin friars, while Niebuhr and Forsskål stayed in the house which was originally meant to accommodate all five men. Although this arrangement was in itself an infringement, however small, of the first rule of the king's 'Instruction', namely that the expedition had to stay together in all circumstances, Niebuhr and Forsskål were glad to have rid themselves temporarily of the philologist, especially as the house had a common kitchen where they had to do their own cooking. Moreover, they expected that Von Haven would soon be removed by royal mandate from the expedition. These hopes were disappointed by a letter from Gähler, which finally reached them towards the end of December.

90. Forsskål to Linnaeus, 20/10/1761, in Fries, Hulth and Uggla, *Bref och skrifvelser*, p. 152.
91. '[...] den oförlikneligaste af all Landskaps-procpect som härifrån ses', Forsskål, *Dagbok*, p. 69.
92. Niebuhr, *Reisebeschreibung*, I.102, I.107.

The Danish envoy had sent a report of their letter of complaint to Bernstorff. In the meantime, he himself was reluctant to believe in such ill-will on the part of Von Haven and advised them to stay united and treat the philologist with respect and confidence. He also warned that a disruption at this stage of the prestigious scientific expedition, which the whole world of learning was following, would displease the king of Denmark. This was an unexpected decision, and rather than facing the dangers of an ill-disposed colleague in the desert, they decided to wait for the ultimate verdict from Denmark in the relative safety of Cairo, where they lived mainly amongst Europeans.

NEXT OPENING
Georg Wilhelm Baurenfeind, engraving of view of Rosetta. Carsten Niebuhr, *Reisebeschreibung nach Arabien und andern umliegenden Ländern*, Copenhagen 1774, plate VI [following page 56]
The Arcadian Library, 12540

Thorkild Hansen has explained the diplomatic details of the months that followed. In his letter to Bernstorff, Gähler played down the amount of arsenic Von Haven had bought and minimized the conflict by insinuating that both the philologist and the naturalist suffered from the 'ancient prejudices prevailing among the two nations in that part of Scandinavia' and that with regard to Forsskål he was 'inclined to believe that he has still retained a good deal of the vulgar character of his home country'.[93] Gähler nevertheless suggested the expedition be split into two parties. Bernstorff, after having consulted Moltke, wrote back to Gähler that he deplored the fact that such an expensive undertaking as this was going to be disrupted because of the passionate temperaments of two of its members, but that he empowered the ambassador to split it up if he thought fit to do so. He warned Gähler that Forsskål was a very ambitious young man, and that, if Von Haven were to remain behind in Cairo and not to carry out his important research on their way to the Yemen, it would make a bad impression if a Swede gained the greatest honours in a project that was meant to give prestige to Denmark. On such premises Gähler ultimately thought it best not to use his delegated powers. He minimized the danger of poisoning, and told the expedition to continue according to the original plan.[94]

It was not in Forsskål's nature to accept such a decision. It was outrageous, he wrote back to Gähler, that 'in order that the reputation of some spineless compatriot might be saved, honest and decent people had been exposed and sacrificed to the most infamous of dangers'.[95]

Gähler's final answer reached the expedition in June 1762, after a stay of six months in Cairo, and almost nine months after they had landed in Egypt. Despite this long period we have remarkably little information about how the members of the expedition spent their time, what they thought, or how they felt. Niebuhr's account of these months, including his elaborate descriptions of Cairo, remains particularly impersonal. The nearest we come to an episode relating how the Europeans encountered the new country is an account of a trip to the pyramids in order to measure their height. While Niebuhr was preparing his astrolabe a young Arab on horseback approached, and asked them who they were and what they were doing. He introduced himself as the son of a sheikh and offered himself as a guide to the pyramids. Despite Niebuhr's and Forsskål's assurances that they did not need a guide, he rode with them. When they passed a small hill, the Arab thrust his spear in the earth and demanded a tip from Forsskål. As the Swede refused to give anything

93. Hansen, *Arabia Felix*, p. 137.
94. Hansen, *Arabia Felix*, pp. 130–43.
95. Hansen, *Arabia Felix*, p. 144.

or to promise anything, Niebuhr rode to the top of the hill and shouted down that there were labourers working in the fields below, upon which the Arab again became quite polite ('höflich').[96]

Hansen surmised[97] that in Cairo the scholars must have felt ever more isolated and estranged from each other. This seems to be borne out by the surviving evidence. During the stay in Egypt Niebuhr travelled with Baurenfeind north to Damietta to explore the eastern branch of the Nile, but Forsskål is not mentioned. Similarly Forsskål went on several excursions around Cairo accompanied only by an Arab servant. If Forsskål and Niebuhr had come to some understanding after their alarming discovery in Rhodes, their friendship had cooled again. Forsskål's relations with Kramer had not improved either. In July he sneeringly wrote once more to Linnaeus about the Dane's 'so-called' knowledge of natural science.[98]

Nor does Forsskål's diary reveal much about the time spent in Egypt. His entry for Cairo consists of a mere six pages, to which is added a long list of the goods and commodities that were transported by the caravans to and from Arabia and sold in the Cairo markets. Moreover, unlike Forsskål's previous entries in his diary, there is hardly any botanical or zoological information in these pages. Although much time had been lost in proceeding to the ultimate destination of the expedition, Forsskål was clearly not idle in these months. He collected plants and animals around Cairo, made a spring excursion back to Rosetta, and repeatedly ventured out into the desert with his Arab servants. These excursions were not without danger, and, as we have already seen, during one of them he was stripped of all his possessions except his trousers. To fend off such dangers, Forsskål let the Arabs who accompanied him collect plants and seeds for him when the country swarmed with Bedouin. He admitted that he did not immediately grasp the full advantage of this device, but once his Arab, after some information and exercise, had become a 'true collecting botanist',[99] Forsskål was able to obtain specimens even in adverse circumstances.

The paucity of botanical names in the diary and letters that record Forsskål's stay in Cairo is misleading. It is explained in the first place by the fact that he was busy sorting out the material he had accumulated so far, and secondly by the fact that he had started describing in a paper the new 'naturalia' amongst the specimens in his collection. So the names and descriptions that once went into his diary were written down separately to be sent off to the world of learning in Europe. By the end of July he was able to tell Linnaeus that he had mailed his new descriptions to Count Moltke and that he had urged him to publish them without delay.[100] Unfortunately, this advice was not followed up.

The material Forsskål had sent to Denmark must have included a list of plants, as well as an elaborate description of the flora of Lower Egypt.[101] These were both published in the *Flora Aegyptiaco-Arabica* of 1775 as 'Flora Aegyptiaca: sive Catalogus Plantarum systematicus Aegypti Inferioris: Alexandriae, Rossettae, Kahirae, Sues', subdivided into three sections with the headings 'Florae Aegyptiacae

96. Niebuhr, *Reisebeschreibung*, I.212.
97. Hansen, *Arabia Felix*, p. 147.
98. Forsskål to Linnaeus, 30/7/1762, in Fries, Hulth and Uggla, *Bref och skrifvelser*, p. 158.
99. Forsskål, *Dagbok*, p. 78: 'efter någon underrättelse och öfning wardt min Arab en hel Botanicus Collector'.
100. Forsskål to Linnaeus, 30/7/1762, in Fries, Hulth and Uggla, *Bref och skrifvelser*, p. 158.
101. Cf. F. N. Hepper and I. Friis, *The Plants of Pehr Forsskål's 'Flora Aegyptiaco-Arabica'*, Kew 1994, p. 29.

idea Geograhico-physica,' 'Plantarum distributio practica' and 'Flora Aegyptiaca'. In the physico-geographical section Forsskål describes the soils, the climate, the habitats, and the geo-botanical communities of the area stretching from the Mediterranean to Cairo and from there to Suez and the Red Sea. This preliminary section is fully justified, Forsskål writes, as the books of the Greeks and Arabs, as well as the more recent publications of travellers, have only covered these facts cursorily and imperfectly.

The Flora section includes 556 numbered species (not counting the cryptogams), of which roughly half are marked with an asterisk and described in the 'Descriptiones Plantarum Florae Aegyptiaco-Arabicae', the final part of *Flora Aegyptiaco-Arabica*, which follows the lists of species found in the Yemen, and which makes up the bulk of Forsskål's book. Niebuhr wrote in the introduction to the *Descriptiones animalium* (1775) that, when he started to work on the material Forsskål had left behind, he had to piece together descriptions written on about 1,800 small scraps of paper.[102] Although Forsskål had considerable time in Cairo to write up and work out the material he had collected—he wrote to Linnaeus that from Egypt he had sent to Denmark a copy ready for the press containing all new naturalia which he had found up to then, together with a flora of Alexandria and a flora of Cairo[103]—there is no way of deciding to what extent the *Flora* as it stands reflects the naturalist's original intentions. There are numerous inconsistencies in the way the work is compiled for publication in 1775,[104] for which it would be unfair to blame Forsskål. It is thus inconceivable that all the species marked with an asterisk were meant as new species, especially as they contain a fair number of names already published by Linnaeus in 1753. But despite the editorial inconsistencies, compared to the 'Flora Constantinopilitana' the lists for Egypt are impressive. About an eighth of all listed species, and about a quarter of marked species, are new. From pp. lix–lxxviii of Forsskål's *Flora Aegyptiaca* it is possible to draw up an imposing list of valid species that can be identified with existing species, and which still carry Forsskål's name to this day.

The botanical results from Egypt were no mean achievement. The relevance of Forsskål's descriptions and naming is that, on the whole, his names are presented according to the Linnaean principles of publication.[105] The majority are specific names of Latin origin, describing the shape of plant parts, while for a handful Forsskål introduced forms that were derived from the Arabic terms he had collected. These include *Antirrhinum haelava*, *Sinapis harra*, and *Genista raetam*.

In 1776 Niebuhr published a volume which was meant to accompany the two works on plants and animals he had published the preceding year. The full title of this book ran: *Icones rerum naturalium, quas in itinere orientali depingi curavit Petrus Forskal*. Its 43 plates were based on drawings made by Baurenfeind during their excursions. Although the book was published with engravings in black and white, two fine hand-coloured copies of the work exist. The first twenty plates of the

102. See Hepper and Friis, *The Plants of Pehr Forsskål's 'Flora Aegyptiaco-Arabica'*, p. 29.
103. Forsskål to Linnaeus, 16/1/1763, in Fries, Hulth and Uggla, *Bref och skrifvelser*, p. 161.
104. See I. Friis, 'Proposals to Amend the International Code of Botanical Nomenclature [ICBN] and Report of the Special Committee on Binary Combinations', *TAXON*, 42, 1992, pp. 343–50.
105. See Hepper and Friis, *The Plants of Pehr Forsskål's 'Flora Aegyptiaco-Arabica'*, pp. 38–42.

Icones depict twenty-seven plants, while the remainder show animals. This is a rather modest number. We do not know how many drawings Baurenfeind made during the expedition, or on what criteria the plates were chosen by Niebuhr. Of Forsskål's list of new species for Egypt only five plants were included. The scarcity of this pictorial material, however, is amply compensated by the wealth of plants that Forsskål collected and pressed, and which safely reached Denmark. They are kept today in the Botanical Museum of the University of Copenhagen.

Engravings of drawings Carsten Niebuhr made of landscapes on the way from Suez to St Catherine's in September 1762.
Carsten Niebuhr, *Reisebeschreibung nach Arabien und andern umliegenden Ländern*, Copenhagen 1774, plate XLIII [following page 232]
The Arcadian Library, 12540

NEXT OPENING
Georg Wilhelm Baurenfeind, engraving of views of Suez and Tur.
Carsten Niebuhr, *Reisebeschreibung nach Arabien und andern umliegenden Ländern*, Copenhagen 1774, plate LI [following page 254]
The Arcadian Library, 12540

The Red Sea

After Gähler's final decision that the members of the expedition should stay together, there was no reason for further delay, especially because, as Niebuhr reminds the reader, the real goal of the journey was Arabia and the southern part of the peninsula.[106]

Christians were not allowed to travel with the pilgrims to Mecca overland. The members of the expedition had to wait till September in order to leave Cairo and travel with a caravan to Suez, from where they intended to sail south to Jiddah. At Suez the party had to stop before continuing their journey on the Red Sea. The Royal Instruction had specifically given Von Haven the task of investigating the inscriptions on Jebel Moqattab on the Sinai peninsula, of providing accurate copies of them and, if possible, of casting the reliefs in plaster.[107] The philologist was also asked to visit St Catherine's, the monastery which was renowned for unique Hebrew manuscripts. The painter had been designated by the king's 'Instruction' to accompany Von Haven on this excursion, but as Baurenfeind had fallen ill with a high fever, and Kramer had to stay with his patient, the philologist was obliged to ask Forsskål and Niebuhr. Forsskål refused, but Niebuhr was willing to go with him.

The excursion, which lasted 18 days, was largely fruitless. They travelled with hostile Arab guides and failed to find any Hebrew inscriptions on the mountain to which they were led. They also failed to get access to St Catherine's, and had to leave without having seen the unique codices in its library. On the way Niebuhr managed to copy some inscriptions on stones of what turned out to be an Egyptian burial place, but otherwise the excursion into the Sinai was disappointing. It was on this part of the expedition that Michaelis had placed high hopes, and Bernstorff could hardly hide his anger when the failure was reported to him. 'All this is utterly deplorable,' he wrote to the members of the expedition, 'This is no way to fulfil the expectations everybody has of our scholars' abilities. It is not for this that all the scholars of Europe have their eyes fixed upon them.'[108] The meagre results of the Sinai excursion severely tainted the prestige of the Danish expedition.

Forsskål, in the meantime, was hardly more successful. During Von Haven and Niebuhr's absence, he did some field work, but because of the season there was little scope for botanizing around Suez. It was as unfruitful, he wrote, as it was dangerous. The risk of being attacked in the open was great and he had to go around armed. The

106. Niebuhr, *Reisebeschreibung*, I.225.
107. 'Instruction' nn. 12, 43.
108. Quoted in Hansen, *Arabia Felix*, p. 187.

Red Sea had more to offer. On the coast he collected shells and described a handful of new species, such as the extremely rare *Malleus regulus*, a rather unattractive member of the Malleidae which is usually found attached to corals. He also went out with a boat to catch fish. These he slit open lengthwise and then, as he did with plants, pressed the skin and part of the skeleton between sheets of paper. He also sought an answer to Michaelis' question about the possible submerged isthmus that would have allowed the Jews to cross the Red Sea during a moment of low tide. This question was particularly dear to Michaelis, as, together with a query about the origin of the name 'Red Sea', it featured at the very beginning of the *Fragen*. Michaelis had studied the problems concerning the division of the waters in detail and ventured a series of hypotheses in the foreword to Pierre Hardy's *Essai physique sur l'heure des marées dans la Mer Rouge comparée avec l'heure du passage des Hébreux*, a treatise republished with Michaelis' observations in 1758. In it he examined the width of the gulf, the location of the place of crossing, the amount of time necessary to reach the opposite bank, and the moment when the crossing was supposed to have taken place. But, as a reviewer of the French Jesuit *Mémoires pour l'Histoire des Sciences et Beaux-Arts* (better known as *Journal de Trévoux*) observed, further information was needed to prove the point beyond doubt.[109] This was exactly what Michaelis had resolved to provide. In formulating his question for the expedition, he suggested that perhaps some natural explanation might be given for the miraculous division of the waters in Exodus 14:21. He wanted to establish whether a north-north-west wind during the ebb tide ('ebb tide upon ebb tide') could lay dry so much of the sea that the Israelites could safely cross to the other side.[110] The place to investigate the exact movement of ebb and flow was the Bedea Valley not far from Suez which, Michaelis maintained, was generally held to be the very spot where Moses crossed the Red Sea. Although it had been ascertained by past scholars that in this far corner of the sea some tidal differences indeed existed, travellers had failed to record the exact times of low and high tide. Michaelis therefore insisted on measuring, on the day before and after the 24th day after a new moon, since he thought 'the crossing must have taken place during one of these two nights'.[111] Apart from calculating the level of the water during low tide, Michaelis also urged Forsskål to inquire of the local fishermen how low the water stood when strong north-west winds drove so much of it towards the Indian Ocean as to create an ebb-on-ebb situation. If such winds could lay bare the bottom of the sea at Bedea, the water must have been quite shallow there and a kind of submerged isthmus must have extended to the other side of the sea to allow Moses to cross with hundreds of thousands of Israelites.

Another objection that had often been advanced against such a natural explanation for the crossing of the Red Sea was that the bed of the sea laid bare in this way would have been too soft and muddy to allow so many people to pass. Michaelis now wanted Forsskål to check whether the bottom of the sea was covered

109. '[I]l manque encoire des observations pour porter au point d'évidence dont il est susceptible', *Journal de Trévoux*, April 1756, Article XLI, p. 968.

110. 'Ihre ganze Absicht ist, mit Gewißheit zu bestimmen, ob durch eine Ebbe über Ebbe, welche ein der Fluth widriger Nord-Nord-West-Wind verursachete, so viel von dem Meer hat ausgetrocknet werden können, daß den Israeliten eine Straße durch dasselbe, geöffnet ward.' Michaelis, *Fragen*, p. 4.

111. '[...] weil meiner Meynung nach in einer dieser zwey Nächte der Durchgang geschehen seyn muß.' Michaelis, *Fragen*, p. 5.

by coral growths which had long since been petrified so as to have been hard and strong enough to carry a man. He also wanted to know what the Arabs called different types of corals, and added: 'I will surely not need to remind the naturalist to pay special attention to the animal which builds and lives in the coral formations.'[112]

While waiting at Suez for Von Haven and Niebuhr to return from their excursion into the Sinai desert, Forsskål set about answering Michaelis' questions. Where the tides were concerned he soon found out from fishermen that ebb at Suez coincided with high tide at Jiddah. The likeliest point for crossing seemed the first narrowing of the gulf near Adabiya just south of Suez. This was, in fact, the place where the Israelites were supposed to have crossed according to tradition. Travelling south by water on 16 September, Forsskål reached Adabiya at sunset. He looked for the monumental stone, and found it was a mere mass of rock without inscriptions. He followed the narrow path through the reedbeds that was supposed to lead to a warm-water spring, but only found a pool with cold water. The endless stretches of reeds at this upper end of the gulf, however, convinced him that the Hebrew name for the Red Sea—Yām-Sūph—which meant 'sea of reeds', derived from this locality.[113] But the people around Adabiya maintained that the historical place of Moses' crossing was near Zafaranah, a day's sailing south of Suez. Moreover, measuring the depth of the sea Forsskål found that the water at Adabiya stood 12 fathoms deep, and that it rose a mere two ells at high tide. Should a double ebb occur, this would only amount to four ells of tidal difference, by no means enough to lay the gulf momentarily dry for the crossing even at Suez, let alone further to the south at Adabiya or Zafaranah. Forsskål was so convinced of this that he did not even bother to venture further south to Zafaranah and take measurements, notwithstanding the fact that he had plenty of time to do so. At this stage he dismissed the whole question in his diary as follows:

The matter belongs even more to the realm of miracle, as the level of the water in times past was higher than it is today. That an extremely low tide should appear at such an important hour, is in itself an exalted matter which savours of a miracle. If one allows for all this, why then hesitate over the obvious conclusion, and straightaway admit that the event was impossible under the common laws of nature, and therefore was a genuine miracle in all its aspects?[114]

Forsskål was clearly more interested in matters of botany and zoology than in biblical questions. After his summary treatment in his diary of Michaelis' ebb hypothesis, he fills pages with descriptions of the merchandise on board the ships sailing the Red Sea. His limited efforts to answer Michaelis' questions must have been just as disappointing to the Danish court as Von Haven's fruitless excursion into the Sinai.

On the afternoon of 9 October 1762 the members of the expedition left Suez on

112. 'Dem Physico brauche ich nicht ins Gedächtniß zu bringen, was ihm ohne mein Erinnern beyfallen wird, auf das Thier genau Acht zu geben, so die Corallen bauet und bewohnet.' Michaelis, *Fragen*, p. 8.

113. Forsskål, *Dagbok*, p. 92.

114. 'Och blifver saken så myket mer återständ till sitt värde af Undervärk, som Vatnets högd i de tider varit ansenligen större än i dag. At en ovanlig Ebb skulle just på så vigtig stund infinna sig och så myket uträtta, är redan en hög sak och i all likhet af under; medgifver man så långt, hvarföre dröjer man med den lätta utlåtelsen, at straxt bekänna händelsen för en omöjelighet efter naturens vanlige Lagar; och altså egenteligen ett under, i sin hela inrättning?' Forsskål, *Dagbok*, p. 93.

board a vessel carrying Mecca pilgrims to Jiddah. The ship made little progress as it sailed very cautiously through coral formations along the coast, and they dropped anchor as soon as night fell. When, on the second day, they reached Zafaranah, Forsskål roused himself a moment to measure the depth of the sea but he found that the water near a sandbank half a mile from the coast was still 13 fathoms deep, so that it was impossible to conceive how any ebb, even ebb on ebb, could have laid the sea dry. Moreover, he wondered, how a could low tide there withstand the weight of all the water of the adjoining ocean for so long?[115] They now crossed to the east coast and stopped at the harbour of Tur for a provision of fresh water. The southern part of the town was inhabited by Greeks, and Forsskål had an opportunity to go on land and visit, with a Greek officer, some of the palm plantations of the area. He also managed to visit ancient Raithu, which some said was the Biblical Elim. He quickly inspected the warm springs at Moses' Baths, and beyond the Monastery of the Valley he found some brackish water inhabited by small fish, which he had no time to examine or describe. He walked back to Tur along the beach and collected several rare shells and seaweeds as well as a *Medusa andromeda*.

The voyage to Jiddah continued, and on 15 October the vessel left the Gulf of Suez, passed Tiran and continued down the Red Sea. On the following day they saw the last mountains of Egypt fading from their view. The coral banks now made it difficult to land, and it was not until four days later that Forsskål had an opportunity to do so again. They slowly proceeded to Jiddah where they finally dropped anchor on 29 October. They rented a house in the town where they stayed till mid-December when they found a ship—and favourable winds—to take them further south to Luhayyah.

Forsskål's diary contains no information about the six weeks' stay at Jiddah. Niebuhr's *Reisebeschreibung*, however, expresses the sense of relief they all felt after the hostility they had met with in Egypt: 'We have nowhere been so apprehensive of the inhabitants of a town as in Jiddah. In Egypt Europeans are approached with so much disdain, that we believed the Muslim feelings against Christians would only increase as we neared the city they hold holy [...] We found, however, that we were completely wrong in this.'[116] In Jiddah they could move freely among the Arabs, visit the coffee houses, and go for excursions. They soon made the acquaintance of a local dignitary, who had a taste for astronomy and natural history. As he only spoke Arab, they all had a good opportunity to practice the language and obtain valuable information about the country. Forsskål visited him frequently, and soon persuaded him to lay out a small garden and send an order to Medina for a Meccan Balsam Tree.

While the expedition was windbound at Jiddah Forsskål did little botanizing and came up with very few specimens. The 'Flora Arabico-Yemen' contains a mere handful of entries of plants for the area: *Salsola vermiculata*, an unidentified species, *S. mucronata*, *Suaeda fruticosa*, both already described at Alexandria, *Dolichos faba* ssp. *nigrita*, an unidentified plant already described for Cairo, and an orache resembling

115. Forsskål, *Dagbok*, p. 103.

116. 'Wir haben uns nirgends vor den Einwohnern einer Stadt so sehr gefürchtet als zu Dsjidda. Weil man den Europäern in Ägypten mit so vieler Verachtung begegnet, so glaubten wir, daß der Eifer der Mohammedaner gegen die Christen sich immer mehren würde, je mehr wir uns den von ihnen für heilig gehaltenen Städten näherten.' Niebuhr, *Reisebeschreibung*, I.279.

Atriplex hastata, but which was rightly described as the new species *A. farinosa.*

In his account of the Danish expedition Hansen marvels at the complete silence in Forsskål's diary about their stay at Jiddah and wonders whether 'the energetic professor [was] suffering from fatigue'. Having no further evidence he writes that 'we must be content to suppose that he accompanied Niebuhr when, every day as evening fell, the latter left the stone house by the harbour for a trip to the quayside.'[117] Admittedly, the number of species collected is extremely low compared to the wealth of species collected elsewhere, but to conclude from this that Forsskål was idle most of his time at Jiddah is misleading. The very meagre botanical results indicate that not much grew in this barren area and that the time of year was inauspicious. To Linnaeus he reported that 'in Suez and Jiddah I obtained next to nothing in plants. All the land around these places is waste, bare, and unusable. The lands are a little green in times of rain, but none fell when I was there.'[118]

Rather than collecting plants and seeds, the naturalist devoted his energy to the fauna of the Red Sea. He studied and described the fish he managed to catch, as well as those that fishermen brought ashore in their nets. The fish were again prepared for preservation by slicing them open and pressing the skin with half of the skeleton between sheets of paper, as he had done at Suez. One of the advantages of this technique is that colour patterns remain visible. Of the 72 species thus preserved in the 'Herbarium Ichthyologicum Forsskåli' in Copenhagen almost half are marked Jiddah, while other species are specimens caught along the coast between Jiddah and Luhayyah. Similarly, approximately 50 per cent of the 'pisces' Forsskål described in the *Descriptiones*[119] are new species (98), half of which are labelled 'Jiddah'. The lists include some species from the Nile, an occasional specimen from Turkey, and many others simply labelled 'Red Sea'. Forsskål's new species include a number of *Carangidae* (jacks). Of these the Golden Trevally *(Gnathodon speciosus)*, a big-eyed shiny golden fish with black stripes, is today a favourite aquarium species. The *Descriptiones* also includes two new halfbeaks (*Hemiramphidae*), fish with a pointed elongated lower jaw: the Yellowtip Halfbeak *(Hemiramphus marginatus)*, and the Blackbarred Halfbeak (*H. far),* a species which Forsskål described from a specimen caught off the Yemenite coast, but which since the opening of the Suez Canal can also be found in the eastern Mediterranean. Forsskål further described three new butterfly fishes (*Chaetodon*), four parrot fishes (*Scarus*), four snappers (*Lutjanus*), five groupers (*Epinephelus*), and his description of the Marbled Spinefoot constituted the new genus *Siganus*. It would be beyond the scope of this study to mention all Forsskål's new species,[120] but it is clear that he wasted no time at Jiddah and that his contribution to the knowledge of the Red Sea fish is considerable. The only two species of fish, both new, depicted in the *Icones*—the elegant Tiera Batfish (*Platax teira*) and the striking Bluespine Unicorn Fish (*Naso unicornis*)—are amongst the most beautiful plates of the *Icones*.

117. Hansen, *Arabia Felix*, p. 204.

118. 'I Suez och Gidda har jag nästan intet wunnit i botaniquen. All mark deromkring är öde, naken, obrukad. Marken grönskar litet i regntiden, men den föll ej in i min tid.' Forsskål to Linnaeus, 16/1/1763, in Fries, Hulth and Uggla, *Bref och skrifvelser*, p. 159.

119. Forsskål placed some of the genera under the heading 'Amphibia'.

120. But see W. N. Eschmeyer's *Catalog of Fishes*, an internet database published by the California Academy of Sciences (www.calacademy.org/research/ichthyology/catalog).

Forsskål also devoted much of his time at Jiddah to the study of corals, as Michaelis had insisted he should do. It is reasonable to suppose that most of the specimens in Forsskål's collection came from the reefs between Tur and Qunfudhah. In the *Descriptiones* Forsskål specifies the fact that between these two towns the coralliferous formations below the surface of the water hills defend the shores, while these formations become rarer further south, or, as far as he knows, might not exist there at all. The study of the coral ridges, of course, made Forsskål reconsider the question of the crossing of the Red Sea, but he only found negative evidence. The location of Moses' crossing, he writes, is not to be sought in any of these parts, as the coral ridges form walls that would have made an ascent from the bottom of the sea impossible. To answer Michaelis' question about a possible coral isthmus, he adds that no solid coral formations can be found in the sea near Suez.[121]

Only 26 corals are described in the *Descriptiones*. Although they include new species, this is hardly a spectacular result. Cyril Crossland, who spent years living in these parts cultivating pearl shell for the Sudan Government, was much disappointed by this number. Admittedly, Forsskål was the first to describe *Porites solida*, one of the commonest species forming massive colonies in the Red Sea, but 'how could anyone go out onto a Red Sea reef,' he wonders, 'and see only one species of *Montipora*, and that one usually not at all the commonest?' Why are there numerous specimens of the same species, while many common species are missing? How could Forsskål obtain 'of the rare *Synarea undulata*, which I have never found myself, three [specimens]'? He deduces from this that 'Forsskål did not collect from the reefs himself.' This procedure corresponds to his employing natives to collect plants for him in Egypt. Crossland further explains: 'his native collectors, as has been my own experience, puzzled beyond words as to what he could want, finding one sort acceptable, sought for more of the same.'[122] It is likely that he obtained shells from the reefs in the same way. However, none of the new species described in the *Descriptiones* are labelled Jiddah.

On 13 December the expedition went on board a small open coffee ship, which, as it was fitted together without nails, looked to Niebuhr more like a barrel than a ship.[123] The crew did not inspire much confidence either. The captain was scantily dressed and looked wild with just a loincloth and his long curved knife. The rest of the crew were black slaves. But as the Europeans were not very likely to find other opportunities to sail south, they decided, not without trepidation, to sail with this ship to Luhayyah in the north of the Yemen.

They made very slow progress, and Forsskål could get on land every now and then. At Ghorab he found numerous shells and collected two plants. On 21 December he explored Qunfudhah,[124] a 'miserable town with crooked streets

Georg Wilhelm Baurenfeind, illustration of *Chaetodon unicornis* in Linnaeus' personal copy of Forsskål's *Icones*. *Icones rerum naturalium: quas in itinere Orientali depingi curavit Petrus Forskål*. Edidit Carsten Niebuhr, Copenhagen 1776, plate XXIII
Linnean Society Library, BL.690

121. Pehr Forsskål, *Descriptiones animalium: avium, amphibiorum, piscium, insectorum, vermium; quae in itinere Orientali observavit Petrus Forskål. Post mortem auctoris edidit C. Niebuhr. Adjuncta est Materia Medica Kahirina atque tabula Maris Rubri geographica*, Copenhagen 1775, p. xxix.

122. Cyril Crossland, *On Forskål's Collection of Corals in the Zoological Museum of Copenhagen*, Spolia Zoologica Musei Hauniensis I, Copenhagen 1941, p. 17.

123. Niebuhr, *Reisebeschreibung*, I.290.

124. It is not clear when Forsskål visited Qunfudhah. The entries Oct. 1762 on p. 229 and p. 235 in Hepper and Friis, *The Plants of Pehr Forsskål's 'Flora Aegyptiaco-Arabica'* are very doubtful. November is more likely.

Georg Wilhelm Baurenfeind, illustration of *Salicornia perfoliata* in Linnaeus' personal copy of Forsskål's *Icones*. *Icones rerum naturalium : quas in itinere Orientali depingi curavit Petrus Forskål*. Edidit Carsten Niebuhr, Copenhagen 1776, plate I
Linnean Society Library, BL.690

NEXT TWO OPENINGS
Georg Wilhelm Baurenfeind, engraving of views of Luhayyah and Bayt al-Faqih.
Carsten Niebuhr, *Reisebeschreibung nach Arabien und andern umliegenden Ländern*, Copenhagen 1774, plate LXI [following page 312]
The Arcadian Library, 12540

and without walls'[125] some 300 kilometres south of Jiddah, but again found very little to rouse his interest. The fact that these places produced very modest results indicates that, as in Jiddah, the area was too barren at this time of year to produce much of interest. In his *Flora* Forsskål consequently lists a mere nine species, none of which carry the collector's name today except for one new plant he had found in great quantities—'copiose ad littora'—both at Jiddah and at Qunfudhah. This was *Salicornia perfoliata* (=*Halopeplis perfoliata* (Forssk.) Bunge). It was drawn by Baurenfeind and included in the *Icones* as the first plate of the volume. Otherwise, Forsskål contribution to the flora of present-day Saudi Arabia remains very modest.

The Yemen

In Luhayyah the expedition was welcomed by the local emir. He was a true, justice-loving gentleman who felt friendly to strangers, Niebuhr wrote, and, when he discovered that they were not traders but scientists, he asked them to meet him and entertain him for a couple of days. If they did, he would put some of his camels at their disposal so that they could travel to Mocha overland. Moreover, a prominent tradesman of the place, who needed a doctor, came forward and offered them hospitality in one of his houses. Niebuhr could only conclude that the behaviour of Muslims improved the further they travelled away from Egypt.[126]

The Emir of Luhayyah was much interested in the scientific instruments which they carried with them. To impress his host, Forsskål offered some money for the first man who would bring him a louse, which he showed him through his magnifying glass. When the rumour spread among the inhabitants that the foreigners were willing to pay money for common lice, Forsskål put this to good use and soon had boys bringing all kinds of beetles and other insects to him. As soon as he found that he could move about unmolested, he started to go on botanical excursions in the surroundings. With relief he wrote to Linnaeus that finally one could botanize in safety even in isolated waste lands.[127] But around Luhayyah gardens were 'pauci & pauperrimi' and on trips to nearby localities he found only few plants of interest in irrigated fields, as the rest was 'desertus, nudus'. An excursion to Mawr, about two hours inland, was more productive: Forsskål listed 78 wild plants, most of which he had not seen in Egypt.[128]

The expedition left Luhayyah on 20 February. Although they decided to travel by land via Bayt al-Faqih to Mocha, the naturalia they had collected were forwarded by sea. For the first time they could travel unescorted with their donkeys, and the Emir of Luhayyah had sent a messenger ahead to make sure that they would meet with hospitality in the villages on the way. They reached Bayt al-Faqih on 25 February, and decided to make it their base from which to explore the north of the country. The members of the expedition now mainly went their own way: Niebuhr rode over the countryside measuring geographical positions for the first detailed

125. 'En ömkelig Stad. Utan mur, krokige gator'. Forsskål, *Dagbok*, p. 115.
126. Niebuhr, *Reisebeschreibung*, p. 297.
127. Forsskål to Linnaeus, 16/1/1763, in Fries, Hulth and Uggla, *Bref och skrifvelser*, p. 160.
128. Forsskål, *Flora Aegyptiaco-Arabica*, LXXXVIII–LXXXIX.

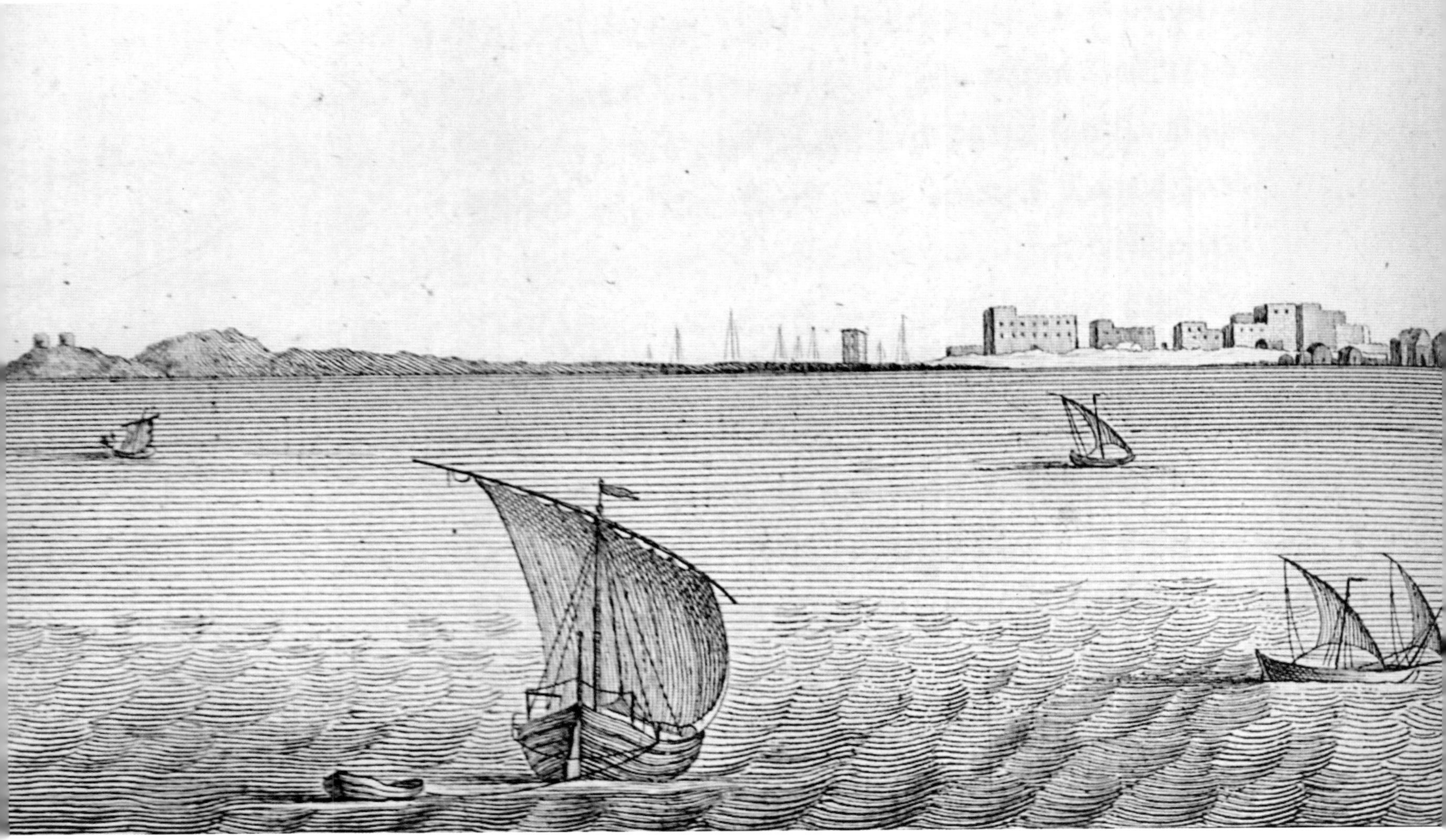

map of the Yemen, and Forsskål went out by himself to study the flora of the fertile Coffee Mountains east of the town. Although the natives travelled by night at this time of year to escape the heat in the desert plain of Tihama, in order to do their research Niebuhr and Forsskål had to travel in the heat of the day, but both were so intrigued by their tasks that they were willing to overlook this inconvenience. Forsskål, in particular, waxed so enthusiastic about the wealth of plants he found in the mountains that Kramer and Baurenfeind decided to join him. So did Niebuhr on 21 March, when he felt he had finished mapping the burning sandy plains. Niebuhr now wanted to start collecting information about the mountainous area for his map, and asked Forsskål to explore the area in more detail with him. To Niebuhr the dialect of the Arabs of the mountainous area was still unknown. He therefore tried to persuade Forsskål to travel with him, to bear him company, and to help him with the language, as he had already learnt in the Coffee Mountains many of the words which were not used in Tihama.[129] Forsskål, of course, realized the advantage of researching the area so systematically and accepted. On 26 February they set out. Two donkeys were rented and they asked their owners to follow them as guides. They had long gone about in Arab dress, but they now also decided to assume Arabic names so that nobody, not even their guides, could guess they were Europeans, and would rather think they were eastern Christians. They often had to sleep in miserable coffee huts or in the open. The food, mainly bad durra bread and some camel milk, was poor, but the fresh water of the mountains, after the brackish fluid of the desert, was a delight.

On 4 April, while crossing the isolated hills south of Mount Sudan along a bad road full of bends, Forsskål noticed a balsam tree in full flower. After some study he believed he had found the true Meccan Balsam Tree, and he felt jubilant with this discovery. In the shade of the tree he carefully made a description.[130] He took some of the flowers with him and showed them to the local inhabitants. They told him that twigs of the tree were used for their delicate fragrance but that they did not know how to produce balsam from it. They called it *ibn sham*, 'the perfuming'. Back in Bayt al-Faqih Forsskål immediately wrote to Linnaeus to announce his find, and promised to write again later from Mocha with more details. The flora of the Yemen had been more rewarding than he had expected, and the Meccan Balsam Tree crowned his achievements. He was in high spirits when he added in his letter to Linnaeus that the country was indeed well worth a botanical exploration and that Professor Michaelis should forever be honoured for having set up such a project.[131] Although Forsskål worked tirelessly from January through to March, there are some signs that by the middle of April fatigue and a longing to go home were beginning to set in. He was not as ardent about his collecting as before, and, in a curious juxtaposition of themes, he writes in the postscript to his letter to Linnaeus dated 18 April that 'from my father I have not received a single letter since I sailed from Denmark. Every now and then I still collect insects, but have not managed

129. Niebuhr, *Reisebeschreibung*, I.337.

130. See Forsskål, *Flora Aegyptiaco-Arabica*, p. 79. Forsskål gave it the name *Amyris opobalsamum*. Today, however, it is known as *Commiphora gileadensis* (L.) Christensen. Forsskål had sent some of its seeds to Linnaeus, who published the new species in 1764, long before Niebuhr returned to Europe with Forsskål's notes.

131. Forsskål to Linnaeus, 18/4/1763, in Fries, Hulth and Uggla, *Bref och skrifvelser*, p. 163.

to describe many.' He also had the feeling that, with the discovery of the Meccan Balsam Tree, the greater part of his botanical mission was over. He added: 'we reached the Yemen from Jiddah on 29 December 1762, and I believe I have already seen and described most of the country's plants, so that I do not have to remain here much longer to supply the rest.'[132]

Although Niebuhr and Forsskål thought they had explored the area around Bayt al-Faqih sufficiently, departure for the south had to be delayed as they had found Von Haven 'etwas unpäßlich' ('a little unwell') upon their return. Niebuhr too had contracted a fever.[133] On 20 April they again felt strong enough to travel, and left for Zabid. To avoid the extreme heat of the Tihama plain they travelled by night, which, of course, meant that neither Niebuhr could continue work on his map, nor Forsskål on his flora. Tantalized by the unseen novelties, the two men therefore decided to move by day, while the rest of the company travelled half a day behind them.

They thus reached Mocha, where they found awaiting them at the customs office the chests they had forwarded from Jiddah. When they went to reclaim their luggage, however, they were in for a nasty surprise. The welcome they received was by no means as hearty as they had experienced in other parts of the Yemen, and when the customs came upon a glass bottle with fish in spirits, disaster struck. Forsskål warned the officers not to open it, as its contents would surely offend their noses. This only made them more suspicious of its contents, and one of the officers promptly opened it, took out some fish and stirred the contents with a rod. The customs house soon filled with the smell of spirits and of rotting fish. 'It is easy to imagine, what the Arabs, whose religion strongly forbade alcoholic drinks, said and how ashamed we must have felt that the Dola and his secretaries saw their customs house so polluted by our possessions,' Niebuhr commented.[134] The officers now also came upon some of Forsskål's sea animals which, not having completely dried, added to the stench in the room. 'This caused further murmurs and curses on the Franks.' Some of the carefully packed shells were mercilessly strewn over the floor, while other packets were pierced by a pointed iron. As the Arabs believed that no normal person would collect such things, many of them had started to think that the Europeans had sent these goods to Mocha merely to play a trick on the Dola and his officers, while others held that some precious wares must be hidden among their possessions. Their amazement knew no bounds when they came upon a flask with snakes in spirits, and someone suggested that these Europeans had come to poison the Muslims. Although the Dola seemed to look upon them more with pity than with scorn, he decided that the foreigners were not going stay the night in the town. It was only by presenting the Dola days later with a nice sum of money that they were finally able to win some of his trust, and to have their damaged possessions retuned to them.

At Mocha Von Haven's health, which had remained weak, deteriorated. He soon felt so weak that he needed help to move from the cool of the terrace, where

NEXT OPENING
Georg Wilhelm Baurenfeind, engraving of village in the Yemeni Coffee Mountains. Carsten Niebuhr, *Reisebeschreibung nach Arabien und andern umliegenden Ländern*, Copenhagen 1774, plate LXIII [following page 334]
The Arcadian Library, 12540

132. 'Från min Far har jag ej fådt ett enda bref, sedan jag rest från Dannmark. Insecter samlar jag nog wid tillfälle, men hinner ej beskrifwa många [...] Till Jemen ankommo wi från Gidda d. 29 Dec. 1762, och jag tror mig redan sedt och beskrifwit landets mästa örter, så att jag ej behöfwer blifwa här så altförlänge att supplera resten.' Forsskål to Linnaeus, 18/4/1763, in Fries, Hulth and Uggla, *Bref och skrifvelser*, p. 164.

133. Niebuhr, *Reisebeschreibung*, I.346.

134. Niebuhr, *Reisebeschreibung*, I.356.

he slept in the open air, to the rooms of the house they had rented. He died on 25 May, presumably of malaria. In Forsskål's diary there is no reference to Von Haven's death and no hint of how he felt about it. Only in a letter to Linnaeus written two weeks later did he mention in passing his relief: 'One of our company, Professor Von Haven, died here on 25 May. His death made the journey much easier for the rest of us. His temperament was rather hard to bear.'[135] It is possible that Von Haven's death contributed to a sense that it was time to return to Denmark, and, since some English ships lay anchored at Mocha, it was decided to travel with them to Bombay, and from there back to Europe. As the ships were not going to sail until the middle of August, however, there was some time left to explore the interior of the Yemen. But as they had met with so much sorrow in Mocha already, Niebuhr wrote, they were not attracted by the idea of using the town as a base for excursions inland.[136] They therefore left for the capital Sana'a on 9 June.

Although some further new plants were found by Forsskål on their journey from Mocha to Ta'izz, by June the season for botanizing in the lower parts was clearly over. The first day they travelled through dry and uninhabited country. On the following days the country became mountainous and more promising, even if in the surroundings the vegetation consisted mainly of succulents. If Forsskål had written to Linnaeus from Mocha that he had already found more than 600 plants in the Yemen, many of which he thought constituted new genera and species, on the way to Sana'a he was able to add only a handful of species to his collection. On the fourth day they reached Ta'izz, where they were received by the local Dola, who had received information about the travellers from the Dola at Mocha. Above the town there were tantalizing glimpses of the fertile Sabr mountains, where according to Arab legend all the plants of the world could be found. Visiting these mountains was something Forsskål had longed for ever since he had heard people talk about them in the Yemen, but now that he was within reach he was not given permission to go there by the Dola. With much difficulty he managed to get a promise instead for an excursion on Mount Sawraq, but this too came to nothing, because marauding tribes made the area particularly unsafe. Forsskål tried again to go to Sabr, and, after much wrangling, permission was given but then again swiftly withdrawn. The expedition met with unwillingness wherever they turned in Ta'izz, and after a frustrating loss of precious time, the men decided on 28 June to proceed on their journey to Sana'a, especially as letters of recommendation to the Imam of the capital had in the meantime reached them from Mocha. On the very day of their departure, Forsskål fell ill. The symptoms were those that had accompanied Von Haven's fatal illness. Everything seemed to be going wrong.

In Niebuhr's travel account the ominous word 'Unpäßlichkeit' appears once more.[137] During the days before leaving for Sana'a Forsskål had not felt well, but neither he, nor the others, were willing to heed the symptoms. After the frustration at Ta'izz, they were only too eager to leave the place, and to explore new territory.

135. 'En af wårt sällskap, Professor von Haven, dog här d. 25 Maj sistledna och gjorde med sin afgång för oss öfriga en ogement lättare resa. Hans sinnelag war rätt beswärligt.' Forsskål to Linnaeus, 9/6/1763, in Fries, Hulth and Uggla, *Bref och skrifvelser*, p. 167.

136. Niebuhr, *Reisebeschreibung*, I.364.

137. Niebuhr, *Reisebeschreibung*, I.391.

But over the following days Forsskål grew so weak that on 4 July the men decided stay at Manzil. Niebuhr too had contracted a slight temperature. However, the camel drivers objected that the place was not suitable for their animals and that the town of Yarim was near. They also promised that they could easily engage people to carry Forsskål there. The following day Niebuhr and Baurenfeind travelled ahead to escape the heat, while Kramer and Berggren stayed behind to travel with Forsskål, awaiting the men who would carry the invalid uphill. On the way Niebuhr had to vomit violently, and when they were all reunited at Yarim he found out that either nobody had been engaged to carry Forsskål or that no Arab had been willing to carry a Christian. They had bound Forsskål on a sickbed on top of a camel and had thus proceeded to Yarim. He reached the place in a wretched state. It was only now, Niebuhr laments, that they realized how miserable it is to fall ill in this country, without having access to some European food or drink.[138]

In Yarim the men rented a house. An initial moment of improvement in Forsskål's condition was soon followed by a final deterioration, and on 11 July 1763 he died at the age of 31. After burying their companion, the four remaining men travelled on to Sana'a, where they met with more hospitality than they had been led to expect. Their minds, however, were by now bent on returning to Mocha and leaving with the English ships for Bombay. It was above all Forsskål, Niebuhr wrote, who had made the Danish expedition a success. With his death the expedition not only lost their best naturalist, but also their best linguist, who had often acted as their spokesman.

Forsskål's explorations in the Yemen were mainly botanical, and they enabled him to draw up a 'Flora Arabico-Yemen' of 626 species. Most of these were collected during the excursions into the coffee hills behind Bayt al-Faqih, and Forsskål must have been able to sort them out to some extent during the remainder of his stay at this town, and later while they were held up at Mocha. We may safely conclude that most of this material was supervised by the naturalist himself.[139] Still, notwithstanding the fact that Forsskål's life was cut short, and with it the exploration of the Yemen, he collected a remarkable number of plants. Of the 626 numbered species in his Flora, about 60 per cent are marked as new. Of these about 40 per cent are today recognized as valid entities. This makes 25 per cent of all species found (approximately 150) new species. Some species he indicated as new were indeed new at the time, but were again described by Linnaeus during the 1760s, long before Niebuhr managed to publish Forsskål's results in 1775. For others Forsskål, or Niebuhr's anonymous editor, failed to provide a description of a new genus, while he did give descriptions of the species. This is the case of the four species of *Saelanthus* (today *Cissus*), which were truly new species of a hitherto unknown genus, and were drawn by Baurenfeind, and published in the *Icones* in 1776.

From the moment of Von Haven's death Forsskål seems to have reckoned with the possibility that he too might not survive. He had already written to Linnaeus twice about his most important discovery, the real identity of the Meccan Balsam Tree. On the very day of leaving Mocha he decided to write once more, and

NEXT OPENING
Georg Wilhelm Baurenfeind, illustration of *Saelanthus quadragonus* in Linnaeus' personal copy of Forsskål's *Icones*.
Icones rerum naturalium : quas in itinere Orientali depingi curavit Petrus Forskål. Edidit Carsten Niebuhr, Copenhagen, 1776, plate II
Linnean Society Library, BL.690

138. Niebuhr, *Reisebeschreibung*, I.395.
139. For the specimens collected during their trip from Mocha to Sana'a, Niebuhr (and his unknown collaborator) must be held responsible for sorting the material.

Tab. II.

included a cutting with leaves and flowers. The letter reached Linnaeus in 1764. 'Yesterday I got a letter from the realm of death,' he acknowledged to a friend.[140] The remaining members of the expedition fled the Arabian peninsula on board an English ship to India, but Baurenfeind died at sea and Kramer in Bombay. The only surviving member, Carsten Niebuhr, travelled back to Europe via Muscat, Baghdad, Aleppo and Istanbul, reaching Denmark in 1767.

140. 'I går fick jag bref från de dödas rike af sal.' Linnaeus to Wargentin, 1764, in Fries, Hulth and Uggla, *Bref och skrifvelser*, p. 169.

CHAPTER 2

THE NATURALISTS OF THE FRENCH EXPEDITION OF 1798–1801

Paris

In 1769, following the Treaty of Paris, when France lost most of its North American territories and its Indian colonies to Great Britain, the French war minister Étienne-François de Choiseul urged that, if France wanted to remain an international trading power, the moment was ideal for a shift of its commercial interests towards the Near East. For such a policy to succeed, he pointed out, the acquisition of Egypt would be a necessary prerequisite. Moreover, with the Ottoman Empire in decline, such a move would be relatively easy. As the French ambassador at Istanbul Count de Saint-Priest was to put it: 'Egypt is at our doorstep. Egypt no longer belongs to the Turks; the pasha is nothing; Egypt does not belong to anybody.' Although the project was hotly debated for years in political circles, nothing came of it at the time. This was mainly because Choiseul was dismissed from office in 1770 and because his successor, Charles Gravier Vergennes, wanted to remain loyal to the Sublime Porte of the Ottoman Empire, under whose domain Egypt officially fell. But the naval minister Sartine disagreed with Vergennes's reluctance to break the peace with Turkey and, in 1777, ordered Baron de Tott on a secret mission to Egypt to study the possibilities of an invasion. A decade later Count de Volney, who published in 1787 his successful *Voyage en Syrie et en Égypte* after travelling to the Near East, harshly condemned Vergennes's policy ('une politique mal-habile').[1] Although he did not approve of western interference in the East, he still emphasized the ideological need to liberate Egypt from the Ottoman Empire. His view that Egypt would benefit from the scientific, economic and cultural progress of the European Enlightenment, however, appealed to many and seemed to justify the very intervention he condemned. But the outbreak of the French Revolution made a military expedition impossible.

In the aftermath of the Reign of Terror plans for conquering Egypt were revived by two very different, but influential, men. In July 1797 Charles-Maurice de Talleyrand read a paper before the Institut de France on the advantages to be obtained from new French colonies, entitled *Essai sur les avantages à retirer de colonies nouvelles dans les circonstances présentes*. In it he recalled Choiseul's aims.[2] He argued that both French industry and culture would benefit from colonial expansion. It would, he added, offer unlimited opportunity for French enterprise and ingenuity, and give space to the many 'active minds in need of projects'.[3] But above all, he

1. Quoted in Henry Laurens, *Les origines intellectuelles de l'expédition d'Égypte: l'orientalisme Islamisant en France (1698–1798)*, Istanbul and Paris 1987, p. 176.

2. In *Essai sur les avantages à retirer de colonies nouvelles dans les circonstances présentes*, Paris 1797, Talleyrand introduces de Choiseul as 'M. le duc de Choiseul, un des hommes de notre siècle qui a eu le plus d'avenir dans l'esprit, qui déjà en 1769 prévoyait la séparation de l'Amérique de l'Angleterre et craignoit le partage de la Pologne, cherchoit dès cette époque à préparer par des négociations la cession de l'Égypte à la France, pour se trouver prêt à remplacer par les mêmes productions et par un commerce plus étendu, les colonies américaines le jour où elles nous échapperoient' (p. 14).

3. '[H]ommes agités qui ont besoin de projets' (*Colonies nouvelles*, p. 16). Also compare J. F. Bernard, *Talleyrand: A Biography*, London 1973, pp. 174–5. Talleyrand's concluding lines read as follow: '[one impetus behind the search for new colonies would be] la douceur de pouvoir attacher à ces enterprises tant d'hommes agités qui ont

stressed the need to prepare for the replacement of their present colonies in order not to be behind the times.[4] Talleyrand's views on British colonial power in the New World and his solution for France added to his growing political prestige, and only two weeks after his paper on colonial expansion he was appointed foreign minister.

In October 1797 general Napoleon Bonaparte, after conquering and subduing Italy, seemed one of Talleyrand's 'hommes agités' eager to find new employment. He wrote to Talleyrand to propose a military expedition to Egypt, during which he would first try to secure Malta and Corfu for France. This tallied with the foreign minister's plans, and Talleyrand raised the issue with the Directors, who, however, neither agreed nor disagreed. Bonaparte and Talleyrand did not give up their plan for Egypt and discussed it over the following months. On 5 March 1798 Talleyrand formally proposed an invasion of Egypt to the Directors, stressing that Russia or Austria would surely take possession of it if France did not. To lighten the responsibility of taking such a momentous decision he added that, although Egypt was part of the Ottoman Empire, Turkey had no real control over the area, and that the Mamluk beys in Egypt had harassed French merchants to such an extent that intervention was overdue. But most convincing of all was his appeal to wider commercial interests. As he had said to Bonaparte the previous year, Egypt, as a colony, would soon replace the products of the Antilles, and, as a route, would open trade with India, making five voyages possible in the time it now took to make three.[5] This time the Directors were convinced by Talleyrand's arguments and realized that for the control of the trade route to the East, Egypt was of central importance. They may also have liked the idea of having Bonaparte out of the way, since his growing popularity was causing anxiety. It was therefore decided that in May a fleet under his command would set sail on a campaign that, in Bonaparte's own words, 'would take anything from six months to six years'.[6]

The historian of the origins of the Egyptian expedition, François Charles-Roux, has summarized the main reasons in favour of the decision for an Egyptian campaign: the dismemberment of the Ottoman Empire, a balance of power in the Mediterranean, French economic interests in the Levant, exploitation of the trade with India, and an attack on English colonial power and hegemony at sea.[7] Such political reasons, together with Napoleon's unbridled enthusiasm at the prospect of military feats during the expedition—'Here all has been done, I cannot find glory any more; this little Europe does not offer enough. One needs to go to the East, all great glories come from there [...] All the East needs is a man'[8]—tend to obscure

besoin de projets, tant d'hommes malheureux qui ont besoin d'espérance.' Talleyrand, *Colonies nouvelles*, p. 16.

4. '[L]e besoin de préparer le remplacement de nos colonies actuelles pour ne pas nous trouver en arrière des événemens.' Talleyrand, *Colonies nouvelles*, p. 16.

5. 'L'Égypte, comme colonie, remplacerait bientôt les produits des Antilles et, comme chemin, nous donnerait le commerce de l'Inde; car tout, matière de commerce, réside dans le temps, et le temps nous donnerait cinq voyages contre trois par la route ordinaire', quoted in François Charles-Roux, *Les origines de l'expédition d'Égypte*, Paris 1910, pp. 303–4.

6. Bernard, *Talleyrand*, p. 202.

7. Charles-Roux, *Les origines de l'expédition d'Égypte*, pp. 339–40.

8. 'Tout s'use ici, je n'ai déjà plus de gloire; cette petite Europe n'en fournit pas assez. Il faut aller en Orient, toutes les grandes gloires viennent de là. [...] L'Orient n'attend qu'un homme,' Napoleon, 25 January 1798, quoted in André Malraux, *Vie de Napoléon par lui-même*, [Paris] 1930, p. 55.

how he first envisaged and later enforced the colonization of Egypt. Charles-Roux's analysis fails to take into account a feature that lent long-lasting grandeur to the French exploits in Egypt: namely the quest for scientific knowledge of the natural and cultural resources of the invaded country, ardently discussed at the Institut d'Égypte and subsequently published in the momentous *Description de l'Égypte*. When Napoleon set out for Egypt on 19 May 1798 with an imposing army of 50,000 men he had more than a thousand civilians with him, including some 120 first-rate technicians (civil engineers and surveyors), scientists and artists. Admittedly, the Egyptian campaign would ultimately prove a military (and political) failure, but the scientific success of the expedition cannot easily be dismissed.

Although Napoleon's mandate from the Directory placed its emphasis on the conquest of Malta and Egypt in order to lessen significantly British political power in the area by dominating the trade route, Napoleon was also told to open the sea route to the Far East by piercing the Isthmus at Suez and to improve the living conditions of the Egyptians in general. One can see how he needed engineers for the first task, and economists, surgeons, engineers, interpreters and printers for the second. To Louis-Antoine Fauvelet de Bourienne Napoleon reported that he would colonize Egypt and that for this reason he would bring 'artists and artisans of every description; women, actors, etc.' with him.[9] Although he did not explicitly mention scientists to Bourienne, they were present in his plans of colonization from the very beginning. Bourienne, trying to give Napoleon all the credit for setting up the expedition, wrote that it was in the end 'Bonaparte [who] organised the army of the East, raised money, and collected ships; and it was he who conceived the happy idea of joining to the expedition men distinguished in science and art.'[10]

Bonaparte's interest in science is well known. The poet Népomucène Lemercier reported him to have said that, had he not become commander-in-chief, he would have dedicated himself to science and followed in the footsteps of the Galileos and Newtons. Because he always succeeded in his great enterprises he would have distinguished himself with his scientific work. No other glory, he concluded, could have tempted his ambition.[11] At the age of fifteen Bonaparte dreamt of a scientific career, and although scientific works featured little on his early reading lists—Dhombres estimated that of approximately 500 hundred pages of notes only about 30 had some scientific flavour—there is no doubt that he always sustained an interest in scientific progress. This is reflected in his expressions of regret when the frivolous fashions of Paris started to re-emerge during the mid-1790s at the cost of the Enlightenment cult of science and learning. He wrote, for example, to his brother Joseph that a huge crowd turned up for a performance in honour of an old actress, the carriages lining up in dream-like procession, but that people were no longer interested in libraries, or courses of history, chemistry, mathematics, literature,

9. Louis Antoine Fauvelet De Bourrienne, *Complete Memoirs of Napoleon Bonaparte*, ed. R. W. Phipps, London 1891, ch. xii.

10. Bourienne, *Complete Memoirs of Napoleon Bonaparte*, ch. xii.

11. 'Si je n'étais pas devenu général en chef ... je me serais jeté dans l'étude des sciences exactes. J'aurais fait mon chemin dans la route des Galilée, des Newton. Et puisque j'ai réussi constamment dans mes grandes entreprises, eh bien, je me serais hautement distingué par des travaux scientifiques. J'aurais laissé le souvenir de belles découvertes. Aucune autre gloire n'aurait pu tenter mon ambition.' Quoted in Nicole Dhombres, 'Napoléon et les scientifiques. 1:1779–1798', *Revue du Souvenir Napoléonien*, 430, 1985, p. 2.

botany, astronomy or music.[12] Further evidence of Napoleon's early endeavours to raise the prestige of learning in the eyes of politicians is in a letter to the Directory, in which he strongly recommended the importance of scientists. In 1797 he sent the mathematician Gaspard Monge, member of the Commission des Sciences et des Arts en Italie (which selected the best art treasures for the conquerors to bring to France), along with General Berthier, to Paris for the ratification of the peace treaty of Campoformio. He entreated the Directory to accept with equal distinction the eminent general and the scientist, as both made theirs an illustrious country and rendered the name of France famous. Not only had Monge earned esteem for France by his conduct in Italy, but, as a scientist, he had also revealed many of nature's secrets and destroyed numerous prejudices. 'We must love men of learning and protect science,' he concluded, as new truths and new discoveries would reveal ever more important secrets for the happiness of man.[13]

Although Napoleon ultimately merely remained an enthusiastic and gifted amateur, he fully recognized the role of pure research in science in the progress of agriculture and industry. Napoleon once wrote to the Italian astronomer Barnaba Oriani that, as science honoured the human spirit, so it ought to be especially honoured in free countries.[14] It is attitudes such as these, rather than any scientific achievements, that helped Napoleon to be elected to the Institut National in December 1797 with the help of his friends the scientists Monge, Berthollet and Laplace.

Bonaparte's courtship of science during the years prior to the Egyptian campaign must also be attributed to his personal political agenda. It is significant that, in his praise of science, he evokes two of its greatest names in association with his own potentialities as a scientist: Galileo and Newton. But although he realized that he himself would never reach lasting glory as a scientist, he wanted at least to make sure that, as a patron of science, he would share in the glory of those who did. A France which did not share in the fame of scientific advance would lessen his own fame as a political leader. In his letter to Oriani he added: 'All men of genius, all those who

12. 'Le luxe, les plaisirs et les arts reprennent ici d'une manière étonnante. Hier, l'on a donné Phèdre dans la salle de l'Opéra au profit d'une vieille actrice. Les places avaient triplé parce que c'était une extraordinaire. Eh! bien, il y avait foule à 2 heures après-midi. Les voitures, les élégants reparaissent ou plutôt, ils ne se souviennent que comme d'un songe qu'ils aient jamais envie de briller. Les bibliothèques, les cours d'histoire, de chimie, de mathématiques, de belles-lettres, de botanique, d'astronomie, de musique se saccadent. Ils n'intéressent guère de monde.' Napoleon to his brother Joseph, 12 July 1795, Napoléon Bonaparte, *Correspondance générale*, II*: La campagne d'Égypte et l'avènement 1798–1799*, [Paris] 2005, letter 310.

13. 'Le citoyen Monge, un des membres de la commission des sciences et arts, est célèbre par ses connaissances et son patriotisme. Il a fait estimer les Français par sa conduite en Italie; il a acquis une part distinguée dans mon amitié. Les sciences, qui nous ont révélé tant de secrets, détruit tant de préjugés, sont appelées à nous rendre de plus grands services encore. De nouvelles vérités, de nouvelles découvertes nous révéleront des secrets plus essentiels encore au bonheur des hommes; mais il faut que nous aimions les savants et que nous protégions les sciences.

Accueillez, je vous prie, avec une égale distinction le général distingué et le savant physicien. Tous les deux illustrent la patrie et rendent célèbre le nom français. Il m'est impossible de vous envoyer le traité de paix définitif par deux hommes plus distingués dans un genre différent.' Napoleon to Directoire Exécutif, 18 October 1797, Napoléon Bonaparte, *Correspondance générale*, II, letter 2169. Dhombres misdates this letter as 18 August ('Napoléon et les scientifiques').

14. 'Les sciences, qui honorent l'esprit humain, les arts, qui embellisent la vie et transmettent les grandes actions à la postérité, doivent être spécialement honorés dans les gouvernements libres.' Napoleon to Oriani, 24 May 1796, Napoléon Bonaparte, *Correspondance générale*, I*: Les apprentissages 1784–1797*, [Paris] 2004, letter 627.

have reached a rank of distinction in the republic of letters, are French, no matter in what country they may have been born.'[15] This remark, written to an Italian, indicates that part of Bonaparte's agenda was to annex all human genius to France, because, on the one hand, he believed that the French nation was ultimately superior culturally and scientifically to other nations, and, on the other, because other nations were not to share in his programme of European Enlightenment. It is for this reason that he took scientists and artists with him to Egypt, and that, in later years as ruler of France, he did much to stimulate research by instituting numerous prizes consistently awarded for special achievements in science.[16]

When the Directory accepted with some reluctance Napoleon's proposal to create a 'commission des sciences et arts' for his Egyptian campaign on 15 January 1798, not much time remained to find scholars and scientists of excellence who were willing to embark on such an adventure. Monge and Claude-Louis Berthollet, one of France's prominent chemists, took most of the recruiting in hand. The country's most important scientific institutions were appraised by them for young scientists or promising pupils. From the École Polytechnique alone 45 men were invited to join the expedition to the East, many of them mathematicians, land surveyors and engineers, while the Jardin des Plantes (National Museum of Natural History) came up with the zoologist Étienne Geoffroy Saint-Hilaire, the entomologist Jules-César Savigny and the botanists Hypolite Nectoux and Alire Raffeneau-Delile. The botanist Antoine François Ernest Coquebert de Montbret was also invited. Most of these men only enjoyed moderate fame as yet, the exception being the 26-year-old Geoffroy who had already made his name four years earlier, with the foundation of the National Museum, when he was appointed professor of zoology. Four of France's greatest naturalists were also urged to join: the mineralogists Déodat de Dolomieu and Pierre Louis Antoine Cordier of the École des Mines, as well as the botanist André Thouin and the palaeontologist Georges Cuvier, both at the Natural History Museum. Thouin and Cuvier declined, but Dolomieu and Cordier accepted. The illustrator of animals at the Museum, Henri-Joseph Redouté, the younger brother of the famous illustrator of roses, was engaged to assist the naturalists in drawing the new specimens they would find.

Toulon and Malta

General Caffarelli du Falga, a man of rare qualities who was very close to Bonaparte, was entrusted with the supervision of the Scientific Committee. As the campaign was to remain secret, the members of the Scientific Committee were not allowed to disclose their destinations. It is hard to say how much the scientists really knew of their final destination. At an early stage Geoffroy communicated to his colleagues

15. 'Tous les hommes de génie, tous ceux qui ont obtenu un rang distingué dans la république des lettres, sont Français, quel que soit le pays qui les ait vus naître.' Napoleon to Oriani, 24 May 1796, Napoléon Bonaparte, *Correspondance générale*, I, letter 627.

16. Raymond Maras concludes that 'rarely has a ruler concerned himself so indefatigably with the patronage of the sciences' (Raymond Maras, 'Napoleon: Patron of Science', *The Historian*, 21, 1958, p. 47); see also Maurice Crosland, *The Society of Arcueil: A View of French Science at the Time of Napoleon I*, Cambridge, Mass. 1967.

at the Natural History Museum that he intended to absent himself for a while as he had received a letter from the government requesting 'ses talents et son zèle' and that he was to leave immediately for Bordeaux.[17] It is not true that Geoffroy's mention of Bordeaux was a ruse of his own, as Hamy maintains,[18] since the destination of Bordeaux was originally given to him by Caffarelli on the orders of Bonaparte. Towards the end of April 1798 Cafarelli asked the scientists and artists who had agreed to take part in the campaign to be on their way. Geoffroy, Savigny and Redouté travelled together to Lyons, where they met the other members of the Scientific Committee, with whom they were to travel to Marseilles and then on to Toulon. Some uncertainty among the scientists as to their destination still persisted. On 5 May Geoffroy, for example, wrote to his colleagues at the Museum that 45 vessels were leaving the port that day for, they believed, Ajaccio,[19] and more than a week later he wrote to Cuvier about the Red Sea and the sands of Egypt, but that the generals had admitted that they were going to India.[20] Most of the members of the Scientific Committee probably knew something about a campaign in the East, and that they would visit part of the Near East in transit, but that Egypt was to be their final destination seems to have remained unclear at this stage. No destination was disclosed to Dolomieu, a Knight of Malta, when Berthollet asked him to join the expedition during one of the meetings of the Académie des Sciences early in 1798. When pressed, Berthollet explained to him that absolute secrecy was required but assured him that there would be plenty of rocks and minerals to study in the country of their destination. When Dolomieu suddenly suspected that Malta was their secret destination, however, he threatened to abandon the expedition. Only when he had Bonaparte's personal word for it that Malta was not their ultimate goal did he decide to join.[21] It may well have been disclosed to him on this occasion that they were to sail for Egypt. Once he discovered he would soon set foot on Egyptian soil, he would have welcomed a means of verifying whether the theory of the formation of the Nile Delta he elaborated in 1793[22]—a remarkable piece of reasoning for a man who had never been to Egypt and had based his conclusions on the observations made by Herodotus and Ptolemy—was valid or not.[23]

17. On 3 April it was read out 'qu'il a reçu du ministre de l'Intérieur une lettre par laquelle le Gouvernement réclame ses talents et son zèle et le charge de se rendre incessamment à Bordeaux.' ('Proceès-verbaux des séances', III.156, quoted by Hamy in Étienne Geoffroy Saint-Hilaire, *Lettres écrites d'Égypte: a Cuvier, Jussieu, Lacépède, Monge, Desgenettes, Redouté Jeune, Norry, etc., aux professeurs du Muséum et a sa famille*, ed. E.-T. Hamy, Paris 1901, p. ii).

18. E.-T. Hamy, in Geoffroy, *Lettres écrites d'Égypte*, p. ii.

19. Geoffroy to the Professors at the Natural History Museum, 5 May 1798, Geoffroy, *Lettres écrites d'Égypte*, p. 15.

20. 'Les généraux ne dissimulent plus que l'on va dans les Indes.' Geoffroy to Cuvier, 13 May 1798, Geoffroy, *Lettres écrites d'Égypte*, pp. 30–1.

21. J. C. Deelman, 'The Life and Times of Commandeur Déodat De Dolomieu (1750–1801)', in *Low-temperature Formation of Dolomite and Magnesite*, Eindhoven 2003, p. 494.

22. 'Mémoire sur la constitution physique de l'Égypte', *Observations et Mémoires sur la Physique*, 43, 1793, pp. 40–61, 108–26, 194–215.

23. Dolomieu's biographer Alfred Lacroix offers two contrasting accounts of Dolomieu's first reaction to the expedition. On the one hand he confirms the complete secrecy of their destination (*Déodat Dolomieu. Membre de l'Institut National (1750–1801). Sa vie aventureuse, sa captivité, ses oeuvres, sa correspondence*, 2 vols., Paris 1921, I.15), while on the other he argues that he accepted because of the possibility it offered him to verify his delta theory (I.52).

Those who had followed the political debate on foreign affairs, however, must have had some inkling about the goal of Bonaparte's campaign. As we have seen, Talleyrand had explicitly referred to Egypt in his *Mémoire*, while many other pamphlets of the time broached the same subject—Delpuech-Comeiras's *Considérations sur la possibilité, l'interêt et les moyens qu'aurait la France de rouvrier l'ancienne route de commerce de l'Inde, accompagnées de recherches sur l'isthme de Suez et sur la jonction de la Mer Rouge à la Méditerranée* (1798), for example.[24]

If the destination was unclear, the possible length and dangers of the expedition were not. When Geoffroy set out for Toulon he felt some qualms about the journey he was about to undertake. He already wrote to Cuvier from Lyons that when he went to see a model of Paris on show at the Palais-Royal of Auxerre he looked tenderly at the Jardin des Plantes and his birthplace, because 'that was the last look for a long time'.[25] And from Marseilles he wrote to his colleagues at the Museum that he dared put trust in their friendship, and that, apart from discussing his scientific discoveries, they would allow him to write about himself from time to time and tell them how he felt.[26]

Geoffroy's reception at Toulon, however, was all that could be hoped for. As might be expected, finding lodgings till the moment of embarkation was difficult, but René-Nicholas Dufriche Desgenettes, the medical commander, offered to share his. There was no need for this. If there had been no distinction between the scientists when they left Paris, Cafarelli now announced that they had all been divided into six different ranks (with respective differentiation in pay), and Geoffroy found himself one of the ten in the upper class. Moreover, at the special request of General Reynier he was asked to sail with him on the frigate *Alceste*, where he was to enjoy the rank of a senior officer. Reynier, Geoffroy wrote to Cuvier, had in the past worked on natural history and wished to spend his spare time on board the *Alceste* discussing science with him.[27] Savigny did not fare as well as his colleague. He was to sail in the *Dubois* and was put into the fourth class only. Geoffroy decided to claim his rights for him.[28]

Cafarelli's division in ranks caused many disputes and complaints among the scientists, as they all claimed to be fit for the first class. Geoffroy commented that especially the less learned had the greatest claims. When they appealed to Bonaparte he decided that all scientists would be downgraded one rank, except the members of the Institut de France. This would exclude Geoffroy, and the members of the Institut protested. To solve the dilemma, Cafarelli now came up with the excuse that Geoffroy was among the names of the persons elected to the Institut, and that as such he had to be admitted to the first rank after all.[29] The zoologist himself did

24. Charles-Roux also mentions a series of other pamphlets on the necessity 'de chasser les Anglais des Indes orientales'. Charles-Roux, *Les origines de l'expédition d'Égypte*, pp. 320–1.
25. 'J'ai cru revoir le Jardin des Plantes et ma patrie; je leur ai donné un regard attendrissant, et c'est le dernier que de longtemps je leur adresserai.' Geoffroy to Cuvier, 29 April 1798, Geoffroy, *Lettres écrites d'Égypte*, p. 3.
26. '[J]'ose espérer de votre amitié [...] que vous n'accueillerez pas uniquement les nouvelles d'histoire naturelle, mais que vous me permettrez de vous parler aussi de temps en temps de moi et de vous informer de ce qui m'arrivera d'heureux ou de fâcheux.' Geoffroy to Cuvier, 5 May 1798, Geoffroy, *Lettres écrites d'Égypte*, p. 15.
27. Geoffroy to Cuvier, 9 May 1798, Geoffroy, *Lettres écrites d'Égypte*, p. 27.
28. Geoffroy to Cuvier, 9 May 1798, Geoffroy, *Lettres écrites d'Égypte*, p. 29.
29. Geoffroy to Cuvier, 18 May 1798, Geoffroy, *Lettres écrites d'Égypte*, pp. 34–5.

not think it necessary to reveal the fact that the Geoffroy Saint-Hilaire on the list was a much older relative of his. Although in the end Bonaparte retracted his decision to downgrade the scientists who had not been elected to the Institut and confirmed Caffarelli's original dispositions, Geoffroy's special treatment indicates that he was considered by Caffarelli, as well as by his older colleagues, as one of the most prominent natural historians of the expedition, and, at the age of 26, was to be ranked with his much older colleagues Desgenettes (36), Dolomieu (48), Berthollet (49) and Monge (52). Savigny, on the other hand, was only 21, Delile 20, and Coquebert merely 18.

The sea voyage was perfectly agreeable to Geoffroy and he wrote that he 'enjoyed all the pleasures of life'. The food was excellent and he thoroughly appreciated the social life on board, conversing with the officers. He wrote to Cuvier that he felt just as happy here as at the Jardin des Plantes.[30] Unlike Forsskål he did not exert himself in the name of science, and apart from dissecting a shark before the officers, hardly any, if any at all, research was done. The other naturalists did not have as much fun as Geoffroy. Redouté found some distraction in observing sea animals, but mostly suffered from boredom. 'Nine days at sea have been enough to make my patience run out,' Coquebert grumbled in a letter to his parents.[31] On 13 June, after the conquest of Malta, the savants finally had an opportunity to leave the ships for a couple of days. Most explored the island's churches and gardens, but no serious research was carried out on its natural history. Geoffroy toured the island on a black donkey, described its geological features as well as its churches, but did not take note, as Forsskål had done, of its flora or fauna.

Setting sail again on 18 June, Napoleon made the expedition's destination officially known. The fleet reached the Egyptian coast and landed in Abuqir Bay on 1 July. After a short but violent battle in the heat of summer Alexandria was conquered on the following day. The members of the Scientific Committee were now assembled on the frigate the *Montenotte* and put ashore. The vessel *Le Patriote*, carrying scientific equipment, ran aground near the entrance of the old port and capsized. Although many of the instruments could be saved, the naturalists lost their herbal paper, spirits and gunpowder.

Alexandria and Rosetta

While the commanders tried to complete the disembarkation as soon as possible, set up a local administration, and assemble the army for the march on Cairo, the members of the Scientific Committee were left to themselves, roofless and without food. Only when Dolomieu complained to Bonaparte that no provisions had been made for them at Alexandria were orders given for a minimum of comfort. Their study of the historical and natural resources of Egypt could now begin.

30. 'Je suis tout aussi à mon aise qu'au jardin des Plantes.' Geoffroy to Cuvier, 23 May 1798, Geoffroy, *Lettres écrites d'Égypte*, pp. 43, 46.

31. 'Neuf jours de navigation ont suffi pour mettre presque à bout ma patience.' Doé de Maindeville, 'La mission d'Ernest Coquebert de Montbret', *Carnet de La Sabretache*, 1956, p. 60; quoted in Yves Laissus, *L'Égypte, une aventure savante avec Bonaparte, Kléber, Menou 1798–1801*, Paris 1998, p. 61.

In the battle to secure Alexandria both General Kléber and General Menou had been wounded. Bonaparte therefore decided that Kléber, who was badly hurt, would stay in Alexandria and made him military governor of the town. His division was placed under the command of Cafarelli, who would march east to conquer Rosetta where, once the town was seized, Menou was made military governor. So the members of the Scientific Committee who stayed in Alexandria fell under Kléber's command. But in July the dry area of Alexandria had little to offer naturalists. When taking possession of Rosetta proved easy, Cafarelli asked Kléber to send most of the scientists over to him. A first group of naturalists, including Geoffroy, Coquebert and Savigny, were all too eager to move on, and set out on 8 July. A second group, including the botanist Hypolite Nectoux, followed a few days later, while by the end of July most of the members of the Scientific Committee had reached the Delta settlement.

In Rosetta life was decidedly more comfortable and interesting for the scientists than in the arid heat of Alexandria. They were given a house, regular rations and four servants, and were soon working on the object of their mission: collecting specimens of the plants and animals of the Nile Delta. Unlike Kléber, who had left the savants to their own lot in Alexandria and had even employed some in clerical and administrative tasks, Menou supervised the scientific work of the Committee. Geoffroy reported that the general was much interested in their success, and in order to procure animal species in the Nile Delta, he gave the naturalists a military escort. Geoffroy was soon immersed in studying a number of unknown bird species. He first studied living species in the wild,[32] and then dissected them for the purpose of anatomical description. He had their skeletons and skins prepared for preservation so that they could be dispatched to Paris. Menou gave orders that all the new species of birds be painted for a projected work on the natural history of Egypt, and he could barely hide his chagrin when he discovered that Redouté, the artist who had come to Egypt for this very purpose, was one of the few who had failed to accompany the naturalists to Rosetta and had remained behind in Alexandria. He had appointed a painter in his stead, Geoffroy warned Redouté, and added that he better join them in Rosetta. Not only could he promise him plenty of milk, grapes and meat, but Menou too had remarked on the artist's absence 'sans humeur'.[33] He also asked Redouté to bring gunpowder and pickling spirits (without which he could not collect and preserve quadrupeds, birds and fish), and to ask Dolomieu to send over some specimens of the rats and gerboas (kangaroo rats) that abounded around Alexandria. The Alexandria rat, which had been mentioned to him by an unnamed naturalist at Rosetta as worth investigating, was caught. Years later a specimen safely reached the Natural History Museum in Paris and was drawn for the *Description de l'Égypte*. Geoffroy named it *Mus alexandrinus*, distinguishing it from the common rat by its characteristic fur, which is brown streaked with gray on the back and uniformly gray on the belly.[34] This rat was among Geoffroy's first zoological results of the scientific

32. It is tempting to read Geoffroy's words 'les observer vivant' (to Jussieu, 12 August 1798) in a modern ethological sense, the word 'ethology' being coined by his son Isidore, who continued much of his father's work, in 1859.

33. Geoffroy to Redouté, 25 July 1798, Geoffroy, *Lettres écrites d'Égypte*, p. 53.

34. Geoffroy further draws attention to the fact that its tail is longer than that of the common rat. See *Description, Histoire naturelle*, II.733–4; illustration, vol. I, plate 5, fig. 1.

expedition. Today the roof rat, as the animal is called in English, is considered a mere subspecies of the common black rat (*Rattus rattus*).

Amongst the first finds of the botanist Raffeneau-Delile around Alexandria was a *Salicornia*, which he discovered growing in great quantities in a ditch near the old port. Redouté drew the plant on the spot, thus justifying his staying behind in Alexandria. When the place was revisited in the following years, the plant was not found again. Delile initially labelled it *S. strobulacea*, but then realized that the plants did not fully answer to the characteristics of this glasswort. He therefore renamed it as the new species *S. nodulosa*. As a result the caption of Redouté's plate and the entry in the *Florae Aegyptiacae illustratio* still have the name *S. strobulacea*, while in the explanatory part of the plates in Delile's *Flore d'Égypte* the new name is published.[35] Delile's name, however, has not been adopted as a valid entity in its own right and is included in *Holocnemum strobilaceum*. Another plant which was drawn by Redouté in Alexandria and included in Delile's description was the succulent *Salsola echinus*. It was found both near the western harbour (Cape of Figs) and on a narrow tract along the coast between the sea and Lake Maryut.[36] Delile maintained that the species was identical to the species Forsskål had found in October 1761 near the catacombs of Alexandria and which he had described as *S. mucronata*. It is odd that Delile should have discarded Forsskål's name and adopted *S. echinus* Labillardière instead.

Although Geoffroy was happy with his numerous new observations in the Rosetta area, the botanists were 'fort malheureux'.[37] The time of year was not propitious for botanical research and they only managed to collect about twenty different species in a month. Seeds were collected and sent to Thouin in Paris, and a number of plants grown from these were included in the *Florae Aegyptiacae illustratio* years later.

If prospects had become more cheerful for the savants after the boredom of the sea voyage and the hardship of Alexandria, their enthusiasm for the expedition was soon dampened by the English attack on the French fleet in Abuqir Bay on 1 August, an event they all witnessed from the rooftops in Rosetta. With the complete destruction of the French fleet they were now cut off from France with no immediate possibility of return. In the meantime Bonaparte had reached Cairo, and, after the Battle of the Pyramids, installed himself in the capital. On 5 August Geoffroy left Rosetta with a small group of savants to assist in setting up the Institut d'Égypte, and within the next two weeks most of the other naturalists followed them to Cairo.

Cairo

'Boït el Cassin-Bey, abou sefre, a rad el Nazari.' Thus Geoffroy heads a letter of 23 August to Cuvier.[38] In Cairo two grand palaces with two adjacent houses were chosen as the seat of the Institut d'Égypte, and the savants and artists were given lodgings there. A vast common room was richly furnished, and the immense park

35. *Description, Histoire naturelle*, vol. II bis, plate 3; *Florae*, II.49; *Flore d'Égypte*, II.147.

36. *Description, Histoire naturelle*, vol. II bis, plate 21; *Flore d'Égypte* II.200.

37. Geoffroy to Jussieu, 12 August 1798, Geoffroy, *Lettres écrites d'Égypte*, pp. 66–7.

38. 'The house of Cassin-Bey, father of sabres, quarter of Nasrieh'; Geoffroy to Cuvier, 23 August 1798, Geoffroy, *Lettres écrites d'Égypte*, p. 73.

around the buildings made an ideal botanical garden. They would soon be better housed than at the Jardin des Plantes, Geoffroy exulted to Cuvier, and he promised to send the first papers presented to the Institut, including two of his own.

Napoleon's arrival in Cairo was marked by preparations for a permanent French occupation of Egypt. The foundation of the Institut d'Égypte is a sign of this, as is the establishment of Joseph-Emmanuel Marc-Aurel's printing press in the capital. Both were part of Napoleon's utilitarian Enlightenment project. Marc-Aurel's *La Décade Égyptienne* particularly reflects the spirit with which the French settled in the capital. In its 'Prospectus' the editor Jean-Lambert Tallien wrote that the conquest of Egypt should not only be seen from a political or economic point of view, but that science and the arts should also benefit from it. The time is past when conquering armies merely destroyed wherever they came, he concluded.[39] Everyday matters and events were published in the daily *Courrier d'Égypte*, so the *Décade* became the vehicle which promulgated the factual information discovered about Egypt. The aim was to get France and Europe to know Egypt, and thus correct both past errors of ignorance and the exaggerated statements made in moments of enthusiasm.[40] As such it reported on the work of the Institut d'Égypte and published its papers.

The first issue of the *Décade* published the proceedings of the first meeting of the Institut of 23 August 1798, during which vice-president Bonaparte put forward a series of questions for its members. Their utilitarian nature is a far cry from Michaelis' elaborated and learned questions for the Danish expedition, and none of them specifically concerned natural history. He, for example, wanted to know if the ovens which the army used to bake bread could be improved, and whether in Egypt a substitute for hops could be found for the brewing of beer. He was likewise interested in finding a way to turn the water of the Nile into drinking water and wondered how it could be chilled. Was Cairo more suitable for watermills or windmills? Were there resources in Egypt for the production of gunpowder? Finally, he wanted to know all about the country's judiciary system, and what changes were generally desired by the population.[41]

At the second meeting Monge read his famous paper on mirages, and Geoffroy was nominated to a committee which was to draw up tables comparing Egyptian and French measures and weights. Furnishing the premises of the Institut with a rich natural history collection was one of the main priorities of the scientists during the first weeks in Cairo, and Geoffroy was able to write to Cuvier on 28 August that their menagerie already contained numerous animals.[42] The environment was also generally conducive to scientific research. Although Egyptian society offered some novelties and curiosities, distraction was limited. After little more than a month Geoffroy reported that he was surrounded by men who only thought of science and that they were all busier than they had ever been in Paris.[43] In the meantime

39. 'La conquête de l'Égypte ne doit pas être utile à la France, seulement sous les rapports politiques ou commerciaux; il faut encore que les sciences et les arts en profitent. Nous ne vivons plus dans ces temps où les conquérans ne savaient que détruire là où ils portaient leurs armes.' *Décade*, I.6.

40. 'Le but que nous nous proposons est de faire connaître l'Égypte [...] à la France et à l'Europe. [...] Ainsi seront rectifiées les erreurs de l'ignorance, les exagérations de l'enthousiasme.' *Décade*, I.6–7.

41. 'Travaux de l'Institut', *Décade*, I.11–12.

42. Geoffroy to Cuvier, 28 August 1798, Geoffroy, *Lettres écrites d'Égypte*, p. 81.

43. 'Je retrouve des hommes qui ne pensent qu'aux sciences je vis au centre d'un foyer ardent de lumières;

Geoffroy had studied ostrich wings and the dog-faced baboon. The results were presented at the meetings of the Institut of 7 September and 27 October. His work on the baboon concerned the identification of the ape which could be encountered in the streets of Cairo and which the ancients had called Cynocephalus. It was the same species, he triumphantly wrote to Cuvier, which Linnaeus had described as *Simia cynocephalus*. Brongniart was therefore wrong to identify the 'Cairo' baboon as the Mandrill, and his 'très habile maître' was wrong to declare the species meaningless.[44] The critical note on the cynocephalus was duly, but briefly, reported on in the fifth issue of the *Décade*. It did not, however, mention Brongniart and Cuvier by name. It merely stated that the animal had been erroneously identified, first by Buffon as the Magot (Barbary Macaque), and later by others as the Mandrill.[45]

Geoffroy's work on ostrich wings was not ground-breaking scientific research: 'You will notice that it was written for the army,' he apologized to Cuvier.[46] But the lack of anatomical detail should not obscure the fact that it was proof of more subtle theoretical thinking. In 1795 he had published for the first time his views on the uniform ground plan of all living beings, a principle which he was to work out more fully two decades later in his famous books *Philosophie anatomique* (1818) and *Histoire naturelle des mammifères* (1819). His argument that vestigial organs in animals indicate common derivation has led some historians to interpret this as a tendency to early evolutionary thinking. Nowhere does such an inclination seem as clear as in the first scientific paper he read to the Institut d'Égypte, where he speaks of the 'modification' of organs. Borrowing the Latin axiom '*natura non facit saltum*,' as would Darwin sixty years later,[47] he observed that 'nature never changes by sudden starts, but that she always leaves a vestige of an organ.' As an interesting further example he says that one can see in the inner corner of the human eye a swollen lump of skin which was a rudimentary form of the nictitating membrane of many quadrupeds and birds.[48] Geoffroy's approach to explaining the anatomy of the ostrich comes surprisingly close to the premises for Darwin's evolution, but one should not exaggerate his importance as an evolutionist. His early interest was archetypal (transcendental) anatomical structure rather than a process of descent from a common ancestor. Still, it was Geoffroy's recognition of homologous structures in animals that would later be used by Darwinists to underpin the principle of evolution. It proved to be a valid method for discerning kinship between species.

During the dry months of August and September there was relatively little work to do for the botanists. This was not so for the zoologist Savigny who made an ample collection of insects.[49] Nevertheless it was he, and not Delile, who read a paper to the

---

nous sommes tous plus occupés qu'à Paris.' Geoffroy to his father, 4 October 1798, Geoffroy, *Lettres écrites d'Égypte*, p. 93.

44. See Geoffroy to Cuvier, 20 October 1798, Geoffroy, *Lettres écrites d'Égypte*, p. 96.

45. *Décade*, I.160.

46. Geoffroy to Cuvier, 20 October 1798, Geoffroy, *Lettres écrites d'Égypte*, pp. 95–6.

47. 'On the theory of natural selection we can clearly understand the full meaning of that old canon in natural history, "Natura non facit saltum".' Charles Darwin, *The Origin of Species by Means of Natural Selection, or the Preservation of Favoured Races in the Struggle for Life*, London 1859, p. 206.

48. '[L]a nature ne marche jamais par sauts rapides, et qu'elle laisse toujours des vestiges d'un organe.' '[S]e voit dans l'homme, à l'angle interne de l'œil, un boursouflement de la peau qu'on reconnaît pour le rudiment de la membrane nictitante dont beaucoup de quadrupèdes et d'oiseaux sont pourvus.' *Décade*, I.50.

49. Geoffroy to Cuvier, 20 October 1798, Geoffroy, *Lettres écrites d'Égypte*, p. 97.

Institut on the discovery of a new species of waterlily. Like other travellers before him—Claude Étienne Savary had waterlilies depicted in his drawings of Egyptian scenes in his *Lettres sur l'Égypte* (1786)—Savigny was impressed by the beauty of stretches of inundated land covered by flowering white and sky-blue waterlilies. He realized that the white form corresponded to the well-known species *Nymphaea lotus* L., but that the blue form did not. Although the Arabs used different names for the two forms—'neoufar' [al-naylufar] and 'bachenin' [al-bashnin]—he found that in European literature only the white form had been described in detail, while there were merely vague and imprecise records for the blue one. Even Forsskål, whom he admired for having produced a very good flora of Egypt, had ignored the plant altogether. His silence about this plant, Savigny adds, is even more surprising seeing that it always grows together with *N. lotus*, 'seemingly finding joy in mingling its flowers with those of the other species'.[50] Linnaeus' latest edition of the *Systema naturae* appeared to have missed this species as well.

Studying the two forms in detail, Savigny found that, apart from the colour of the flowers, the edges of the big round leaves were very different and that the anthers were dissimilar in shape. From this, he argued, one could only conclude that they were two separate entities. He gave the new species the name *Nymphaea caerulea*. The Blue Egyptian Lotus Savigny described turned out to be the flower that figured in the ancient creation myths of Egypt. Its fragrance and supposed narcotic effect have also led some writers to identify it with the flowers of the lotus-eaters in Homer's *Odyssey*. However, when Savigny described the Egyptian species in more detail in the 'Flore d'Égypte', he maintained that Homer's plant was the Libyan suffrutex *Rhamnus lotus*, and did not attribute any narcotic effect to either *Nymphaea lotus* or *Nymphaea caerulea*. He repeated that *N. caerulea* is the flower which is painted ubiquitously on the ancient monuments and he added that the tubers and fruits of both species were still eaten in modern Egypt. A somewhat unattractive uncoloured illustration of both the white and blue species was included amongst the plates of the *Description*.

The description of *Nymphaea caerulea* was Savigny's first important work in Egypt. Of the presentation of this find by the 21-year-old scientist before the Institut d'Égypte Geoffroy wrote to Cuvier that it was written in a stimulating style and that he had acquired a reputation for knowledge which was hard to believe in one so young.[51]

Although Geoffroy was thus highly satisfied with the first botanical and zoological results presented at the Institut, the importance of his work was not clear to the uninitiated. François Bernoyer, a captain of the army who attended the meeting of 7 September, recorded in a letter his impressions of the almost three hours of discussion on whether the ostrich was meant to fly or to run. He was shocked that the discussion reached no conclusion. It was all very 'foolish' and he felt exasperated by the idea that men of learning were wasting their time discussing 'a subject of such insignificance'. He believed that the most ignorant person could have made all

NEXT OPENING
Alire Raffeneau-Delile, plate of *Nymphaea lotus* and *Nymphaea caerulea*.
Étienne Geoffroy Saint-Hilaire, Jules-César Savigny *et al.*, *Description de l'Égypte*, Paris 1809–12, *Histoire naturelle*, vol 2 bis, Botanique, plate 60
British Library, 1899.k.1

50. 'Son silence à cet égard est autant plus étonnant, que la plante dont il s'agit croît toujours à côté du N. lotus, et semble prendre plaisir à confondre ses corolles avec les siennes.' *Décade*, I.71.

51. '[S]on mémoire étoit écrit d'un style très piquant et lui a donné parmi nous une réputation de savoir qu'on refusoit d'abord à son jeune âge.' Geoffroy to Cuvier, 20 October 1798, Geoffroy, *Lettres écrites d'Égypte*, p. 97.

1
a
b

Plée Sc.

the savants agree on the matter by simply telling them that, if nature had given the ostrich long legs, this was for running, and that it would have given it big wings if the animal had been meant to fly.[52]

There seems to have been a decline of interest in the work of the naturalists in the *Décade* too. After the full text of Geoffroy's ostrich paper, the editor Jean-Lambert Tallien only reported very briefly on Geoffroy's discoveries about the baboon and with even greater brevity on a paper he read to the Institute on 1 November.[53] Admittedly, its subject—the periodical shedding of antlers by deer—had little to do with Egypt, but Savigny's paper of 21 November, in which he reported on all the animals observed so far, with special emphasis on his collection of insects, was dismissed in a few lines. On the other hand, much space was given to Antoine-François Andréossy's work on Lake Manzalah, Jean-Joseph Marcel's work on geography and Joseph Regnault's chemical analysis of the water of the Nile. The results of these studies were evidently of a more utilitarian character.

Lake Manzalah

From 21 November 1798 to 4 July 1799 there is no further mention of the naturalists in the *Décade* and thus no indication of the progress they had made in exploring the country. Geoffroy's letter to De Jussieu, director of the Museum in Paris, however, is a perfect, if gloomy, account of what had been achieved by the end of October 1798. His own work on the birds of Egypt had not been very propitious, he wrote, and the collections had not grown as he had hoped. He found that most of the birds of the Delta were the same species as those he had observed in France, and he could do little more than dissect them. Only with snakes and other reptiles had he been more fortunate. Trying to find the species his colleague Lacépède in Paris had described in his voluminous *Histoire des quadrupèdes, ovipares et des serpents* (1788–9) and *Histoire naturelle des reptiles* (1789), he continued to come across new species. But field work was dangerous. Bedouin often lay in ambush behind low dunes, 'pitilessly killing and pillaging'.[54] Geoffroy also had reason to complain about Gérard, a young naturalist whom he had taken with him to skin the animals he caught: the young man, limited by an infantile intelligence, had been of no use at all.[55] Most of the preparations of the animals fell to Geoffroy himself. He intended to send Gérard back to France and asked De Jussieu for another amanuensis. Besides, the botanists had not been able to do much in this season apart from growing plants from the seeds they had collected, and the mineralogists had hardly fared any better. Only Savigny, studying insects, had met with a certain amount of success.

Towards the end of November 1799 all this changed, when most of the naturalists had decided to explore Lake Manzalah and the eastern part of the Nile Delta. In

52. 'En la matière, je crois que le plus ignorant aurait pu mettre nos savants facilement d'accord en leur disant simplement que si la nature avait doté l'autruche de grande jambes, c'était pour courir, alors que pour voler, elle lui aurait donné de grandes ailes.' François Bernoyer, *Avec Bonaparte en Égypte et en Syrie 1798–1800*, Paris 1981, p. 131.

53. *Décade*, I.196.

54. Geoffroy to De Jussieu, undated, Geoffroy, *Lettres écrites d'Égypte*, p. 100.

55. Geoffroy to De Jussieu, undated, Geoffroy, *Lettres écrites d'Égypte,* p. 101.

October the flooding of the Nile Delta reached its highest point, while the water levels started falling rapidly again during the following months. The shallow brackish lagoons and marches in the eastern Delta formed by this process were ideal habitats for birds and fish. In his 'Mémoire sur le lac Menzaléh' General Andréossy, who had explored the geography of the area during the first half of October, had briefly drawn attention to the fact that the lake was 'très-poissonneux' and that bird species abounded on the beachy edges, along the narrow strip of land that separates the lake from the Mediterranean, and especially in those parts where the waters receded after the flooding.[56] This warranted an expedition, and on 23 November a group of scientists including Geoffroy, Delile, Savigny, and Coquebert left Cairo to join Cafarelli's division in Damietta.[57]

Savigny exulted in his diary at the sheer numbers of species of birds in the eastern Delta at this time of year: there were flamingoes, whose hoarse and repeated cries could be heard, as well as pelicans, cormorants and ducks; and, he wrote, during the night one could hear the cries of all these birds together.[58] Unfortunately no letters exist which record Geoffroy's impressions, but it was here that he made the first of a series of discoveries in anatomy that were of great interest for his budding theory of analogies. His attention was drawn to the big mudfish that were common in the lower Nile area. These air-breathing catfish, which can reach a length of one metre and a weight of about seven kilograms, were known to move over land and bury themselves in the mud as the pools were drying up. Dissecting specimens, which he identified as Linnaeus' *Silurus anguillaris*, his attention was drawn to the fact that, apart from gills, the animals also had small rudimentary lungs. When he found a similar species in the upper Nile in 1799, he decided to place both species under the new genus *Heterobranchus*. The description was published by his son in 1809 in the *Description de l'Égypte*, while Redouté's plates had come out the previous year. These plates are amongst the very few in the *Description de l'Égypte* that show the anatomical details of fish.[59]

The mudfish Geoffroy caught in Lake Manzalah is now known as *Clarias anguillaris*. However, his description of a new Upper Nile species remains valid, and he was also right to place it in a new genus. The Upper Nile Mudfish is called *Heterobranchus bidorsalis* to this day. But apart from the question of nomenclature, the air-breathing organ arising from the gill arches was of special interest to him. It formed further proof of homologous structures in different classes of animals. The example of the mudfish is actually used by Geoffroy to sustain one of the central conclusions of his *Philosophie anatomique des organes respiratoires sous le rapport de la détermination et de l'identité de leurs pièces osseuses* (1818)—the book in which he proposed to show 'that there is no part of the bone structure of fish which has not its analogue in other vertebrate animals':[60]

NEXT TWO OPENINGS
Hand-coloured engravings of Spur-winged Plover, White-tailed Plover, Common Redshank, Egyptian Plover. *Description de l'Égypte*, Paris 1808 [proof copy of 1809 edition], *Histoire naturelle*, vol. 1, Zoologie–Oiseaux, plate 6
The Arcadian Library, 09142

56. Andréossy, *Décade*, I.190.
57. Other scientists who joined the excursion were the astronomers Nouet and Méchain, and the geologists Dolomieu and Cordier.
58. Paul Pallary, *Marie-Jules-César Savigny, sa vie et son oeuvre*, Cairo 1931–4, p. 55.
59. *Description, Histoire naturelle*, vol. I, plates 16–17.
60. '[I]l n'est aucune partie de la charpente osseuse des poissons, qui ne retrouve ses analogues dans les autres animaux vertébrés.' Étienne Geoffroy Saint-Hilaire, *Philosophie anatomique des organes respiratoires*, Paris 1818, p. 1.

I have already had occasion to describe [...] a fish which I have brought back from my travels, the silurus anguillaris [...] Behind its ordinary branchiae [...] there are supplementary branchiae [...] One cannot doubt the analogy of those supplementary branchiae to the four arches of branchiae conserved there in the normal state, nor the analogy of these ramifications to the bronchi of the lungs in other vertebrate animals.[61]

Immediately following this reference to the air-breathing mudfish he had studied twenty years before in Egypt, he concludes this section of his treatise by stressing that there exist two systems of respiratory organs (one adapted for breathing in air, the other under water) and that both systems are co-existent in all vertebrate animals according to 'the unity of organic composition for all those beings',[62] even if one of the two systems will necessarily be dominant.

The Natron Lakes

Since ancient times the dry lake beds in the Wadi Natrun in the desert south of Alexandria had yielded a powdery solid known as natron, a substance which was used for the embalming of bodies, as a quickening agent in the process of smelting ore, and in the production of glass. The Old Testament also mentions its medicinal property as an antiseptic (Jeremiah 2:22). Towards the end of January 1799 it was decided that a small group of members of the Institut d'Égypte would follow General Andréossy in a seven-day military survey of the area. Those who were invited to join the expedition included the chemist Claude-Louis Berthollet and the mathematician Joseph Fourier. Redouté was also asked to join. Berthollet and Fourier were already men of great scientific fame, but Andréossy found it necessary to explain to the readers of the *Décade* that Redouté was an able artist who produced coloured drawings of plants, animals, and, above all, of fish.[63]

On 4 February 1899 a report of the expedition was read out before the Institut, and subsequently published in the *Décade*. It deals mainly with the description of the geographical, demographic and physical make-up of the area—the valley which contains the lakes is seen as an ancient part of the Nile valley that once extended as far as Libya—and only a few paragraphs are dedicated to its natural history. Andréossy writes, for example, that on the plateau separating the Nile from the valley of the lakes one can only find a handful of frail plants, 'small, without vigour and very much dispersed throughout the area'. Forsskål's *Hyosciamus datora* is one of them. However, on the edges of the natron lakes one can find rushes in abundance and a few other products of the vegetable kingdom, including a number of plants Forsskål had described as new: *Arundo maxima*, *Tamarix gallica*, *Statice aphylla*, *Fagonia scabra*,

PREVIOUS OPENING
Henri-Joseph Redouté, hand-coloured engraving of a Black and a White Ibis.
Description de l'Égypte, Paris 1808 [proof copy of 1809 edition], *Histoire naturelle*, vol. 1, Zoologie–Oiseaux, plate 7
The Arcadian Library, 09142

Hand-coloured engraving of Spotted Eagle.
Description de l'Égypte, Paris 1808 [proof copy of 1809 edition], *Histoire naturelle*, vol. 1, Zoologie–Oiseaux, plate 1
The Arcadian Library, 09142

NEXT OPENING
Hand-coloured engraving of Black Kite and Pharaoh Eagle-owl (*Bubo ascalaphus* Savigny, 1809).
Description de l'Égypte, Paris 1808 [proof copy of 1809 edition], *Histoire naturelle*, vol. 1, Zoologie–Oiseaux, plate 3
The Arcadian Library, 09142

61. 'J'ai déjà eu occasion de décrire [...] un poisson que j'ai rapporté de mes voyages, le *silurus anguillaris* [...] En arrière des branchies ordinaires [...] existent d'autre branchies supplémentaires [...] On ne saurait douter de l'analogie de ces branchies supplémentaires avec les quatre rangées des branchies conservées là dans l'état ordinaire, comme aussi de l'analogie de ces arbres creux avec les bronches du poumon des autres animaux vertébrés.' Geoffroy, *Philosophie anatomique*, p. 402.

62. Geoffroy, *Philosophie anatomique*, pp. 403–4.

63. 'Habile artiste pour les dessins colorés des plantes, des animaux et sur-tout des poissons.' *Décade*, II.94.

H.N. Zoologie. POISSONS DU NIL. *Par Mr. Geoffroy de Saint Hilaire. Dessinés par H.J. Redouté.*

5

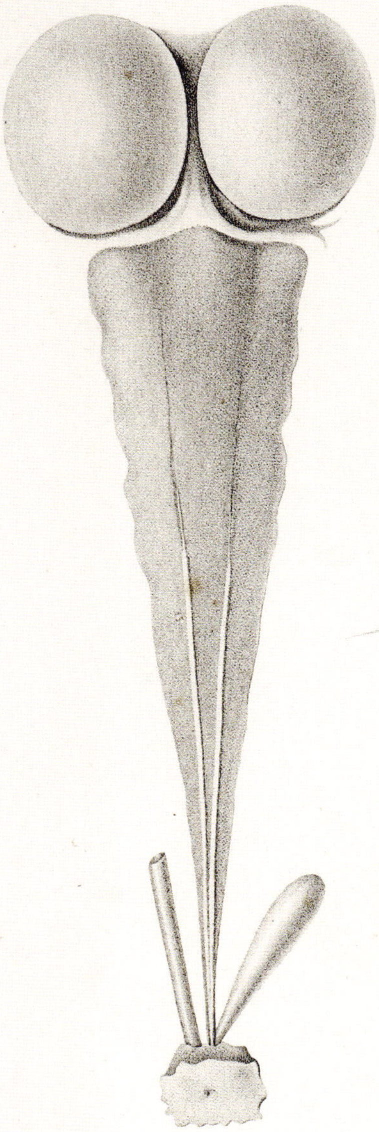

4

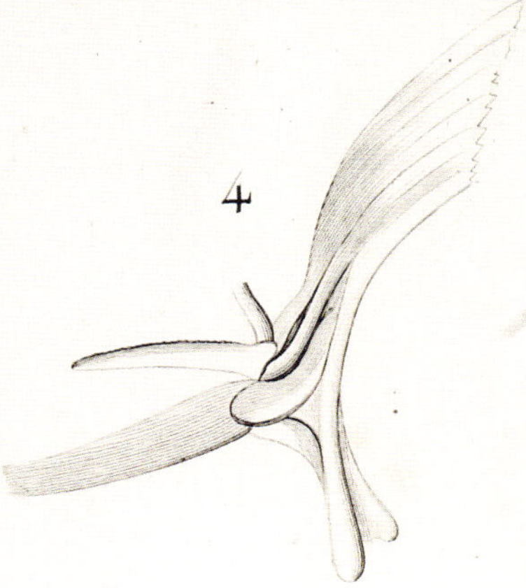

Pl. 16.

1

3

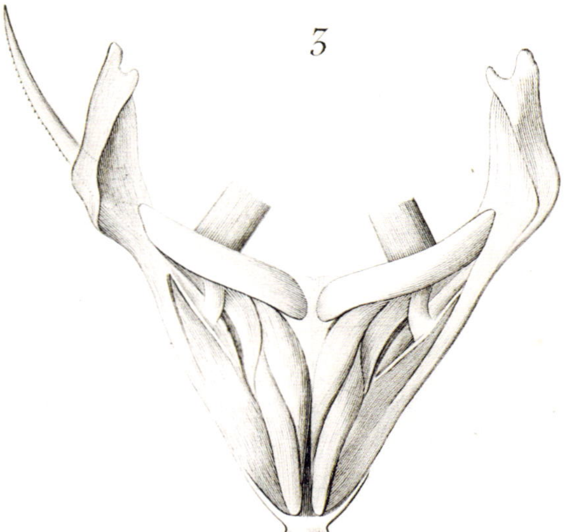

Bouquet Sc.t

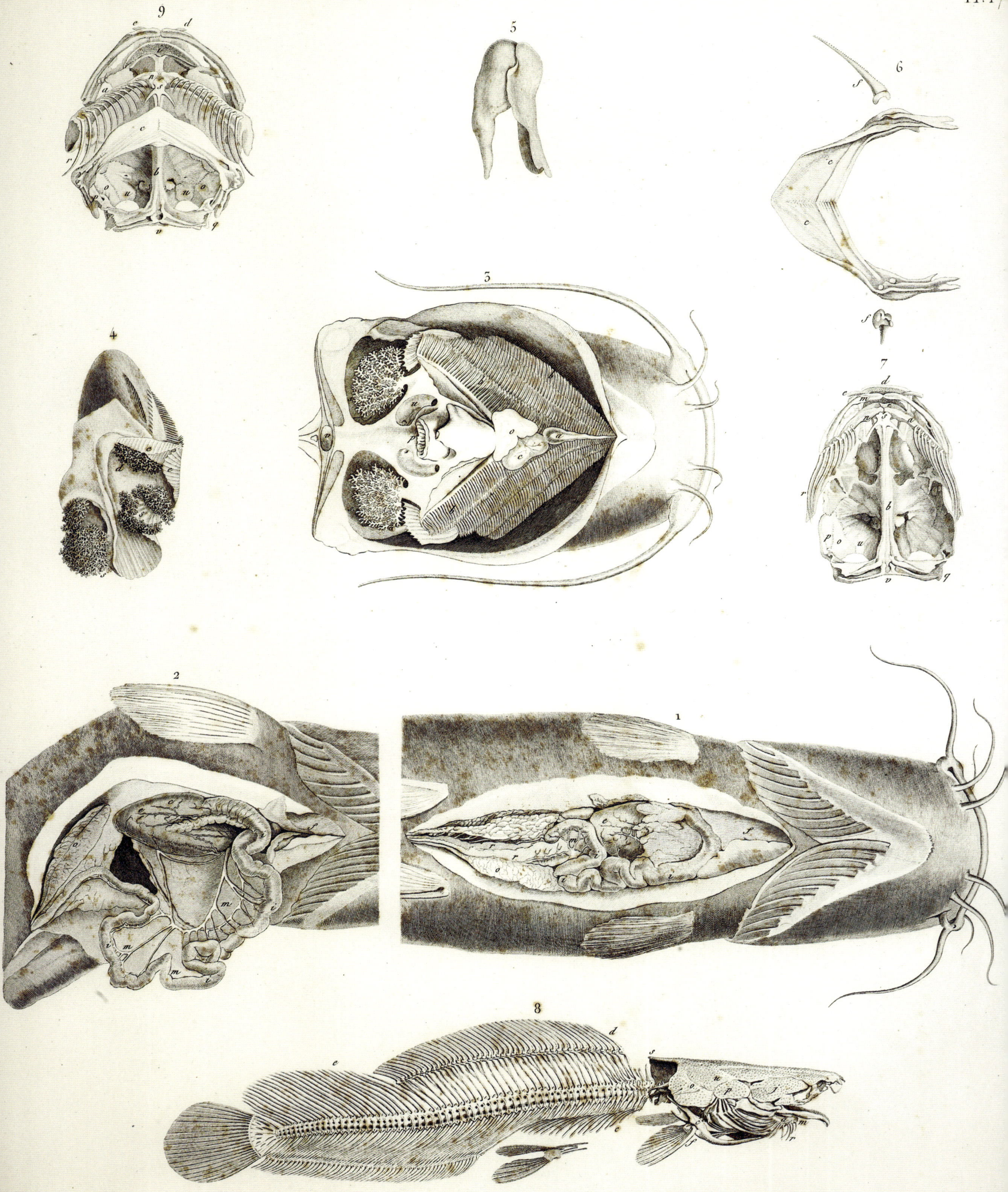

HETEROBRANCHES. *Détails Anatomiques.*

Bouquet

Suaeda vera.[64] A few animal species were observed: two kinds of beetles (*Pimelia muricata, Carabus variegatus*), two ant species, a gnat, a snail, camels and gazelles. At least three species of duck were present in great numbers.[65] As there were no other naturalists on Andréossy's expedition, it seems likely that Redouté was responsible for these records.[66] He might well have collected specimens of plants and animals which were later correctly identified by his colleagues at the Institut and inserted in the published report which came out more than six months later in the fourth issue of the second volume of the *Décade*, dated 'l'an VIII'. As the revolutionary year VIII started on 22 September 1799, the fourth issue must have been published some time during the first months of 1800.

The main object of the excursion to the Natron Lakes was twofold. If Andréossy's task lay in the military recognizance of the area, Berthollet's was to investigate whether soda mining could have any commercial potential for the French. But it was only natural that, once on the spot, Berthollet should want to examine how sodium carbonate was actually formed. It was this research which gave the expedition to the Wadi Natrun scientific importance. Berthollet found that soda is formed in the warm damp soil by the influence of the great quantities of the calcium carbonate of the limestone hills east of the lakes on the sodium chloride of sea salt. Andréossy commented in his report that Berthollet's theory was founded on new observations on affinities.[67] Indeed, Berthollet's observations would lead to the awareness that chemical affinities are influenced by the masses (ratio) of the substances involved. He would work out this important contribution to theoretical chemistry after his return to Paris in 1799 in *Recherche sur les lois des affinités chimiques* (1801), and later in his most important publication, *Essai de statique chimique* in 1803.

The Nile

When Bonaparte started out on his Syrian campaign in February, Geoffroy and a number of savants were asked to follow the army. However, as Geoffroy was recovering from illness and did not feel strong enough to undertake the journey, Savigny went north with the army while Geoffroy returned to Cairo, where he lived serenely ('fort paisiblement').[68] At the slave market he bought an eleven-year-old child for 250 francs who could look after his growing collection and whom he instructed in stuffing animals. There was not much field work that could be done in Cairo, but he found it was well worth looking for rare species brought to the Cairo fish markets. It was here that he came across the Nile Bichir (*Polypterus bichir*). The fish's primitive appearance, with its spiracles and its series of fleshy dorsal pectoral

PREVIOUS OPENING
Henri Joseph Redouté, plate of *Heterobranchus anguillaris* and *Heterobranchus bidorsalis*.
Étienne Geoffroy Saint-Hilaire, Jules-César Savigny *et al.*, *Description de l'Égypte*, Paris 1809–12, *Histoire naturelle*, vol. 1, Zoologie–Poissons du Nil, plate 16
British Library, 1899.k.1

Étienne Geoffroy Saint-Hilaire, anatomical plates of *Heterobranchus*.
Étienne Geoffroy Saint-Hilaire, Jules-César Savigny *et al.*, *Description de l'Égypte*, Paris 1809–12, *Histoire naturelle*, vol. 1, Zoologie–Poissons du Nil, plate 17
British Library, 1899.k.1

NEXT OPENING
Henri-Joseph Redouté, anatomical plate of the Nile Bichir.
Étienne Geoffroy Saint-Hilaire, Jules-César Savigny *et al.*, *Description de l'Égypte*, Paris 1809–12, *Histoire naturelle*, vol. 1, Zoologie–Poissons du Nil, plate 3
British Library, 1899.k.1

64. Only *Suaeda vera* is a recognized species today. It is a pity Andréossy has no further information on the plant he identified as *Arundo maxima* Forssk., the exact nature of Forsskål's species still remaining dubious.

65. Andréossy, *Décade*, II.95, II.102.

66. Cf. Delile's description for *Artemisia monospermia* in the 'Flore d'Égypte': 'elle a été trouvée par M. Redouté sur le chemin de Terrâneh aux lacs de natroun' (*Description, Histoire naturelle*, p. 264.). This find was antecedent to Delile's in the Egarement Valley in the winter of 1799–1800.

67. '[C]ette théorie, que le citoyen *Berthollet* se propose de développer, est fondée sur des observations nouvelles sur les affinités.' *Décade*, II.100.

68. Geoffroy to his father, 23 June 1799, Geoffroy, *Lettres écrites d'Égypte*, p. 20.

H.N. Zoologie. POISSONS DU NIL. *Par M.r Geoffroy de Saint Hilaire. Dessinés par H.J. Redouté.*

1. LE POLYPTÈRE BIC

2. *Sa tête vue en dessus*. 3. *son squelette*. 4. *os de la nageoire*. 5. *les mêmes isolés*. 6. *myologie de la bouc*

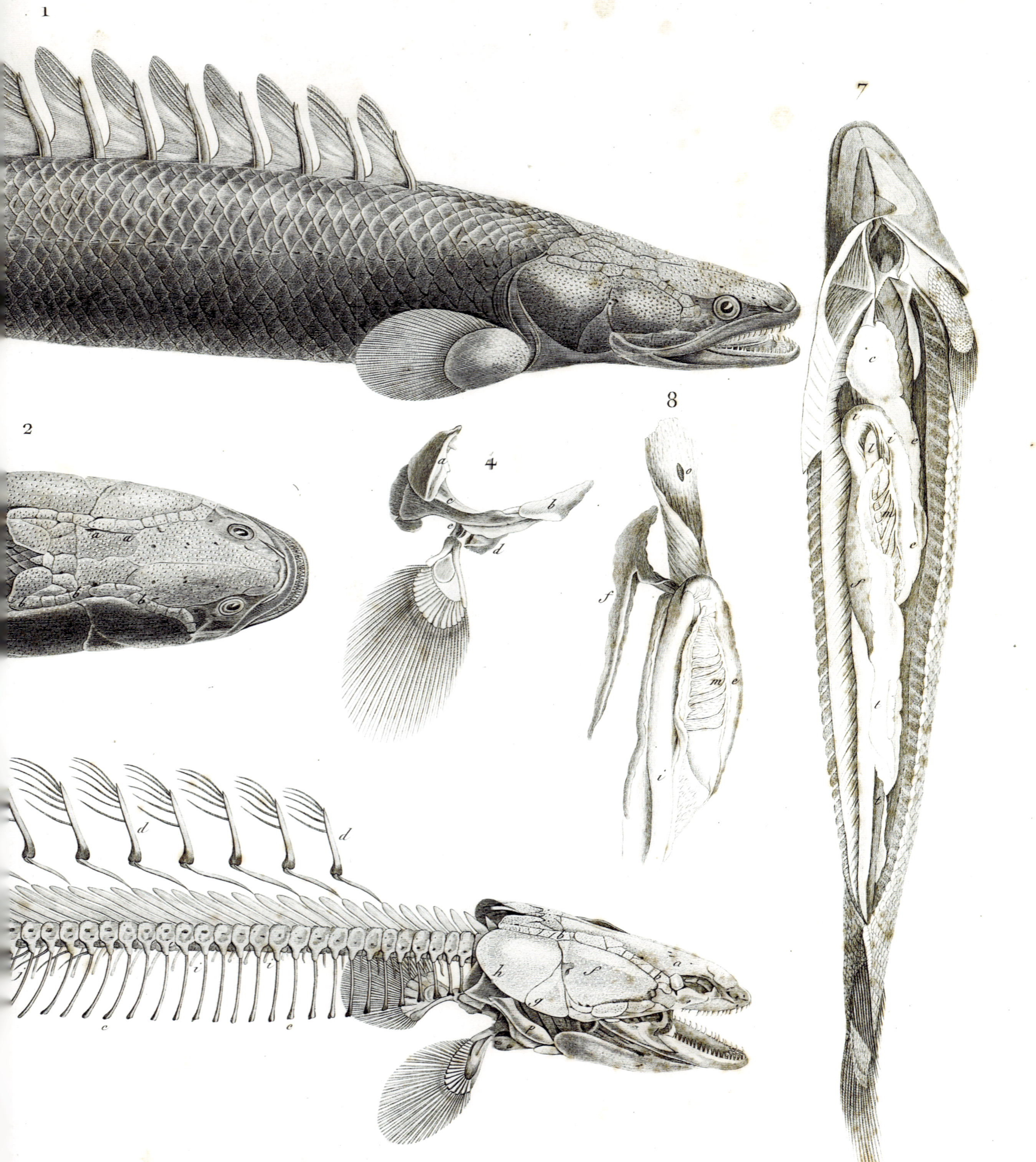

POLYPTERUS BICHIR.

Bouquet Sc.

ères de l'abdomen . 8. l'estomac et le foie . 9. les deux vessies natatoires . 10. la mâchoire inférieure .

fins, required close attention. The animal was no doubt rare. Some of the fishermen he asked declared that they had never caught it before. Geoffroy offered a generous sum of money for further specimens, and the very few that were brought to him proved that the fishermen had told the truth. The mysterious animal seemed to live in the deepest parts of the Nile, and Geoffroy was unable to find out about its habits or what it fed upon. But when he dissected it and discovered that this was yet another fish with rudimentary lungs that allowed it to breathe atmospheric air, and that it used its powerful ventral and pectoral fins for terrestrial locomotion, he realized its importance for his theory. Ten years later in the *Description de l'Égypte* he stated proudly that, if he had discovered only this species in Egypt, that alone would have repaid the fatigue of a long voyage. He did not know of a more remarkable animal. It was the bichir which made him fully appreciate the extent to which nature can deviate from its common types. The animal broadened his ideas on the plan of nature.[69]

As the animal was unique among fish, Geoffroy decided to create a new genus for it. This he did in a publication of 1802, immediately after his return to Paris.[70] However, the vernacular name he gave it—Polyptère (i.e. with many fins)—was not valid, and the authorship of his colleague Lacépède at the Museum, who had duly Latinized it, must now be adopted. The genus is thus known as *Polypterus* Lacépède 1803, while the species is known as *Polypterus bichir* Geoffoy Saint-Hilaire 1802. Today the Nile Bichir is a favourite species for large recreational aquariums. They are aggressive and hardy animals which are known to escape easily. Many collectors have found them out of the water on the floor.

As the waters of the Nile receded, and left pools of water with trapped fish, Geoffroy looked with interest at how children became absorbed in a fish they called *fahaka*, a pufferfish, which, when threatened, inflates itself in a few seconds to several times its normal size. He noted how the children loved to observe them in the pools, predict the moment of their metamorphosis, or, more cruelly, throw them as balls or squeeze them till they burst. When dried in its inflated state with the air entrapped in its belly, the animal kept its spherical form for a long time and could be used as a ball.[71]

The Nile Puffer had been described by Linnaeus' disciple Fredrik Hasselquist with the specific name *fahaka* in his *Iter Palaestinum* (1757),[72] a publication which has, however, no nomenclatural validity (the starting point of zoological nomenclature is 1758 and no name before that date has taxonomical validity).[73] His name has therefore been rejected for *Tetraodon lineatus* Linnaeus 1758. Although the decision to invalidate Hasselquist's work for zoological nomenclature rests on twentieth-

69. 'Je n'aurais découvert en Égypte que cette seule espèce, qu'elle me dédommageroit des peines qu'un voyage de long cours entraîne ordinairement; car je ne connois pas d'animal plus singulier, plus digne de l'attention des naturalistes, et qui, montrant combien la nature peut s'écarter de ses types ordinaires, soit plus susceptible d'agrandir la sphère de nos idées sur l'organisation [...] Le bichir paroît en effet comme un composé d'élémens qu'on ne rencontre que dans les animaux fort différens les uns des autres.' *Description, Histoire naturelle*, I.4.

70. 'Histoire naturelle et description anatomique d'un nouveau genre de poisson du Nil, nommé polyptère', *Annales du Muséum d'Histoire Naturelle*, 1802, pp. 57–68.

71. *Description, Histoire naturelle*, I.22

72. Fredric Hasselquist, *Iter Palaestinum eller Resa til Heliga Landet*, Stockholm 1757, pp. 400–5.

73. Article 3 of the *International Code of Zoological Nomenclature*, 4th edn., 2000.

century ratifications of the International Commission on Zoological Nomenclature, the controversy about the right name of the Nile Puffer goes back a long way, starting with Geoffroy's comments in the *Description*. He starts his section on fish with this species, and heaps abundant praise on Linnaeus' young disciple, while he comments with some asperity (and irony) on the master for changing without sufficient reason an apt Egyptian epithet which 'our traveller had taken upon himself to preserve':

he was soon criticized by his master, who, pervaded by the idea that only characteristic words should be used instead of common names, hastened to replace the term fahaka*, which had no meaning, by the epithet lineatus, which did have an important meaning.* [74]

Geoffroy explains that he agrees with Linnaeus' general principle that names should describe distinct physical characteristics, but that such epithets become inconvenient when further species with similar characteristics are discovered. Why?

When the great German ichthyologist Marcus Elieser Bloch was presented with a striped specimen of a puffer for his monumental *Allgemeine Naturgeschichte der Fische* (1782–95), he gave it the name *lineatus*, implying that he identified it with Hasselquist's species, and produced a magnificent plate depicting a puffer with a striped belly.[75] This was used by later ichthyologists as typical of the Egyptian species. In so doing, Geoffroy complains, they forgot Hasselquist's original description of the species, and missed the fact that the Nile Puffer has a striped back. Geoffroy therefore decided it was time to drop Linnaeus' name in favour of Hasselquist's, because he could no longer use a name which had caused such a confusion of species.[76]

To help clear up the taxonomical confusion around the *fahaka* once and for all, Redouté was asked to produce a good watercolour of the Nile Puffer, which was meant to contrast with Bloch's plate. It was to be the first reliable drawing of the fish. Admittedly, Guillaume Rondelet had produced a simple line engraving of the animal in book XV of his *De piscibus marinis* in 1554, but this was an early pre-Linnaean work. It was no coincidence that Redouté's plate was given some prominence as the very first in the fish section of the *Description de l'Égypte*.

The importance of Redouté's drawing was stressed by the fact that he was asked to present his watercolours officially during the 28th session of the Institut on 4 July 1799, an event which was duly reported in the *Décade*. The paper stressed that the *in-folio* drawings which Redouté had shown during the previous meeting were of interest both for the correctness of their drawing and for the truthfulness of their colours, achieved by copying live individuals (dead specimens could not render such true forms and colours). The article also reported that Redouté had been assisted by Geoffroy who had pointed out to the artist which characteristics were essential

OVERLEAF LEFT
P. Haas, hand-coloured engraving of *Tetraodon lineatus* in Marcus Elieser Bloch, *Allgemeine Naturgeschichte der Fische*, Berlin 1786–7, part II, plate 141
British Library, 954.d.20

Henri-Joseph Redouté, plate of 'Le Tetrodon Fahaka' in Étienne Geoffroy Saint-Hilaire, Jules-César Savigny *et al.*, *Description de l'Égypte*, Paris 1809–1812, Histoire Naturelle, vol. 1, Zoologie–Poissons du Nil, plate 1
British Library, 1899.k.1

OVERLEAF RIGHT
Woodblock print of *Tetraodon lineatus* in Guillaume Rondelet, *De piscibus marinis*, Paris 1554, p. 419
British Library, 461.l.7

74. '[I]l ne tarda pas à en être blâmé par son maître, qui, pénétré de la nécessité de n'employer pour noms triviaux que des mots caractéristiques, s'empressa de remplacer le terme de *fahaka*, qui n'avait aucun sens, par l'épithète de *lineatus*, qui en présentoit un significatif.' *Description, Histoire naturelle*, I.19.

75. M.E. Bloch, *Allgemeine Naturgeschichte der Fische*, Berlin 1782–95, plate 141, opposite p. 129.

76. 'Je ne pouvois plus employer une expression qui avoit produit cette confusion d'espèces.' *Description, Histoire naturelle*, I.20. He further adds that the animal was known to the Greeks under name *physa*, another name, he adds, which is to be preferred to *lineatus*.

foret, nisi eas veteres literis consignassent. Plinius igitur de multis arboribus, auibus, piscibus, cæterisque peregrinis animantibus scripsit, Galenus & Dioscorides de iis quorum magnus est in medicina vsus, quo nomine tantum abest, vt reprehendendi sint, vt maxima iis sit habenda gratia, quòd eorum studio & opera effectum sit, vt multa quę etiam hodie à peregrinis, & procul à nobis remotis regionibus aduehuntur, agnoscamus, magnúmque ex iis fructum percipiamus. Quare nemo iure id mihi vitio verterit, si tantorum virorum exemplo peregrinos pisces à curiosis viris qui eos vel viderunt, vel de iis aliquid scripserunt acceptos bona fide studiosis proponam, vt ij scriptis meis admoniti, si quando inter nauigandum, vel peregrinandum propiùs intuendi & contemplandi occasio se obtulerit, integram de his historiam nobis relinquant: de iis loquor, qui cum summo cognoscendarum rerum studio & curiositate, summam doctrinam ex vetustorum & spectatissimorum autorum libris petitam atque confirmatam coniunxerunt, qua qui carent, non multò doctiores ex longa peregrinatione redeunt. sed persępe in rerum quas viderunt, delectu atque diiudicatione cæcutiunt. Quo fit, vt cùm primùm sibiipsi imposuerint, deinde aliis temere credentibus imponăt. Quod qui seriò docti sunt, facilè perspicient in eius scriptis qui nuper de aquatilibus & aliis quibusdam rebus veluti centones consarcinauit.

De Orbe.

Tetrodon hispid
Linn. 411

CAPUT II.

ORBICVLATOS Pisces vel orbes primùm describemus, quos sic appello, quòd in orbẽ & rotundam figuram circumacti sint. Horum aliquot sunt genera, quorũ primum ex Orientis, alia ex Septentrionis plaga delata vidi. Est igitur orbis primus is, qui à Venetis *pesche colombo*, à Græcis quibusdam huius ætatis *flascopsaro* for distinguishing and classifying individuals.[77] What seemed implicit in this report was that watercolours on vellum from *fresh* specimens of fish under the supervision of the zoologist Geoffroy would put the artist on his guard against the method used by Bloch. These original sheets, most of them now in the Bibliothèque Centrale of the Museum of Natural History in Paris, served as a model for the black and white engravings reproduced in the *Description de l'Égypte*.

Although Geoffroy gives the question of the name *fahaka* much prominence in the opening passages of his essay, the real interest of the species for him lay in the animal's imperfectly understood anatomical structure. By what mechanism did the *fahaka* puff itself up? What was the nature of the bladder-like organ that received the air? How was the air kept inside? Bloch and Lacépède had ventured to suggest that the *fahaka* used a specific receptacle for inflation, but observations based 'on a shrunken, and badly preserved, specimen in spirits' would fail to establish the complete mechanisms involved. Just as Redouté's watercolours of fish were superior to others because they were based on fresh specimens, Geoffroy's anatomical studies on fresh specimens were bound to be more exact. Having all the tetrodons he could wish for 'at his disposal' he opened up a great number of them[78] and convinced himself that the receptacle was no less than the animal's stomach which can expand to over a hundred times its original size. Groups of muscles attached to the oesophagus can both pump the stomach full of air, until the animal becomes almost three times its original size, and also ensure that the air remains blocked inside. Such a defence mechanism of swelling into a much bigger rigid ball that quickly moves to the surface of the water is a remarkable example of evolution. But Geoffroy did not know about evolution. To him, however singular the adaptation of the structure of the pharynx and oesophagus in the *fahaka*, it was not a surprising fact in itself, as he was becoming convinced that 'nature never works through new creations, but only through the modification of already existing forms'.[79] The heterobranchus, the polypterus, and the *fahaka* all proclaimed the same thing. He concludes his essay on the *fahaka* in the *Description de l'Égypte* by saying that it is part of nature's organizing principle that structures which have lost their original purpose can become 'pièces utiles' for other functions.[80] These discoveries made him write triumphantly to Cuvier that 'I pretend to no less than the throne of anatomy [...] have you found in a single fish the plan of quadrupeds and of cuttlefish? Have you explained how the admirable organs of tetrodons work?'[81] To his father he reported with more

77. 'Le citoyen *Redouté* communique à l'Institut une note concernant la peinture des poissons du Nil [...] ; elle ne pouvait être mieux réparée que par les ouvrages du citoyen *Redouté*: ceux qu'il a exposés dans la séance précédente intéressent autant par la correction du dessin que par la vérité du coloris. Les individus ont été copiés vivans, la nature morte ne pourrait fournir des formes aussi pures et des nuances aussi vraies. Le citoyen *Redouté* s'est appliqué, de concert avec le citoyen *Geoffroy*, à représenter fidellement les caractères propres à distinguer et à classer les individus: il s'est assujetti au format *in-folio* de la collection des vélins du muséum d'histoire naturelle, dont ses dessins doivent être une suite.' *Décade*, II.191

78. *Description, Histoire naturelle*, I.23

79. '[L]a nature n'opère jamais par des créations nouvelles, mais simplement par des modifications des choses existantes ailleurs.' *Description, Histoire naturelle*, I.24.

80. *Description, Histoire naturelle*, I.36–7.

81. 'Je n'exige rien moins que le trône anatomique [...] avez-vous trouvé dans un seul poisson l'organisation des quadrupèdes et de la seiche? Avez-vous expliqué comment agissent les organes si admirables des *tetrodons*?' Geoffroy to Cuvier, 16 August 1799, Geoffroy, *Lettres écrites d'Égypte*, p. 129.

modesty that he had already collected enough material for a work of which he expected some success.[82]

Upper Egypt

On 14 June 1799 Bonaparte returned from a disastrous campaign in the Holy Land. Although he defeated the Ottoman forces in some battles, his army was severely crippled by the plague and a lack of supplies, and the stalemate before Acre eventually forced him to return to Egypt without having conquered the Ottoman province of Syria. The campaign was badly planned from beginning to end. Early on the victory at Jaffa had been achieved by butchering over 2,000 helpless Ottoman troops, and the approximately 3,000 prisoners of war, who had surrendered voluntarily, were subsequently massacred. Moreover, the whole campaign cost the lives of thousands of French soldiers, including General Caffarelli. When Bonaparte was forced to lift the siege of Acre towards the end of May, his army, with over 2,000 sick and wounded, was in no condition for a retreat into Egypt and many died on the way. 'The whole track between Acre and Gaza is strewed with the dead bodies of those who have sunk under fatigue, or the effect of slight wounds,' Sidney Smith wrote to Nelson.[83] Not knowing what to do with the plague patients in his field hospitals, Bonaparte ordered them to be poisoned.[84]

It was no wonder that the general mood of the French in Cairo deteriorated. During the session at which Redouté read a paper on his drawings a violent row broke out when Bonaparte proposed a committee to study the plague. Desgenettes, chief medical officer of the Armée d'Orient, refused to partake in such a project and alluded to the poisoning of patients. The assembly became noisy, and Berthollet, who was in the chair, called in vain for order. Murmurs were heard of 'despotisme oriental'. The ideal with which so many had set out, to bring civilization and prosperity to Egypt, had been seriously damaged. It is hardly surprising that Bonaparte took the first opportunity to return to France. He sailed, in secret, with a handful of generals, and with Monge and Berthollet, on 18 August, leaving General Kléber in command of the army. It was a sign for the savants that the Egyptian adventure was drawing to an end.

Although Geoffroy felt he had done enough research, he wanted to visit the Red Sea and Jiddah before returning to France. He wrote to the professors at the Museum on 16 August 1799 that he had long dreamt of travelling to the place where Forsskål had described so many new species.[85] Nothing came of this, however. On 14 August Bonaparte ordered two committees of savants to join the army in Upper Egypt. It was his last initiative in the scientific exploration of the country, but one on the success of which he had set his mind. His last act before leaving was to recommend

82. Geoffroy to his father, 15 August 1799, Geoffroy, *Lettres écrites d'Égypte*, p. 124.
83. Quoted in Christopher Herold, *Bonaparte in Egypt*, London 1963, p. 309.
84. Herold, *Bonaparte in Egypt*, p. 307.
85. 'J'ai songé pendant longtemps à faire le voyage de Djedda; c'est là qu'on trouve un marché de poissons très approvisionné et qu'arrivent les nombreuses espèces que Forskal a décrites.' Geoffroy to the Professors at the Muséum, 16 August 1799, *Lettres écrites d'Égypte,* p. 128.

General Dugua, who was to accompany and protect the savants.[86] Two days after he started north on his return journey to France, the naturalists embarked for the south. In his letter with instructions for the new commander-in-chief, written from Alexandria on 22 August, Bonaparte did not forget to provide for the savants. He stipulated that as soon as their mission in Upper Egypt was completed the committee of arts and science would be put in the hands of a member of parliament in France who would decide on further scientific initiatives.[87]

The discovery and archaeological description of ancient Egypt is usually seen as the greatest and most spectacular scientific achievement of Napoleon's Egyptian campaign. The artist Dominique Vivant Denon had joined General Desaix's division to explore the antiquities of the Upper Nile at the end of 1798. The emotional experience of seeing the marvels of ancient Egypt are recounted first-hand in his *Voyages dans la Basse et la Haute Égypte*. The very thought of reaching the site of Hermopolis (Achmounin), where the ancient buildings had awaited him for four thousand years to show their artistic perfection, made him sigh with happiness.[88] But this was only the beginning. The gate to the temple at Tintyra (Dandarah) made him think that he had come to the very shrine of science and art,[89] and an officer of the army, Major Latournerie, is reported to have exclaimed that ever since he reached Egypt he had been sad and ill, disappointed with everything, but that what he saw at Tintyra had amply rewarded all the suffering: 'whatever will be my lot during the rest of the expedition, I will forever rejoice at the memories of this day.'[90] And this again was only a prelude to Thebes. When the army advanced up the Nile and came suddenly upon the panorama of the ancient city, an 'émotion électrique' ran through the soldiers, and, spontaneously halting, they clapped their hands.[91] It is debatable how long such enthusiasm lasted. Desaix's expedition to Upper Egypt was a prolonged and futile military campaign to rout Mourat-Bey and his Mameluks, and it was full of hardship, disease, bloodshed and cruelty to the local population. But it was with glowing reports and many drawings of ancient Egypt, often made in haste, that Denon returned to Cairo in July 1799, and his paper for the Institut did not fail to make a deep impression. 'If any conquest had been made, the conquest was his, and it was never lost,' the American historian Herold concludes.[92]

Did the naturalists who followed in the artist's footsteps ten months later in August 1799 feel just as enticed by Upper Egypt? Denon, who was not very precise

Dominique Vivant Denon, page with views from Egypt; bottom, the garden of the Institut d'Égypte.
Voyage dans la Basse et la Haute Égypte, pendant les campagnes du general Bonaparte, Paris 1802, plate 25
The Arcadian Library, 08962

86. 'De lever toutes les difficultés qui pourraient se rencontrer, et de faire partir la commission des sciences et arts.' Bonaparte to Dugua, 17 August 1799, letter 4738, Bonaparte, *Correspondance générale*, II.
87. 'La commission des Arts passera en France sur un parlementaire que vous demanderez à cet effet, conformément au cartel d'échange, dans le courant de novembre, immédiatement après qu'il auront achevé leur mission. Ils sont, dans ce moment-ci, occupés à ce qui reste à faire, à visiter la Haute-Égypte.' Bonaparte to Kléber, 22 August 1799, letter 4758, Bonaparte, *Correspondance générale*, II.
88. Vivant Denon, *Voyages dans La Basse et la Haute Égypte, pendant les campagnes de Bonaparte, en 1798 et 1799*, Paris 1802, I.113, p. 147.
89. '[J]'étois réellement dans le sanctuaire des arts et des sciences.' Denon, *Voyages dans La Basse et la Haute Égypte*, I.136–7.
90. 'Depuis que je suis en Égypte, trompé sur tout, j'ai toujours été mélancolique et malade: Tintyra m'a guéri; ce que j'ai vu aujourd'hui m'a payé de toutes mes fatigues; quoi qu'il puisse en être pour moi de la suite de cette expédition, je m'applaudirai toute ma vie de l'avoir faite par le souvenirs que me laissera éternellement cette journée.' Denon, *Voyages dans La Basse et la Haute Égypte*, I.140.
91. Denon, *Voyages dans La Basse et la Haute Égypte*, I.141.
92. Herold, *Bonaparte in Egypt*, p. 262.

Pl. 104.

I

Denon Del & Sculp.

1

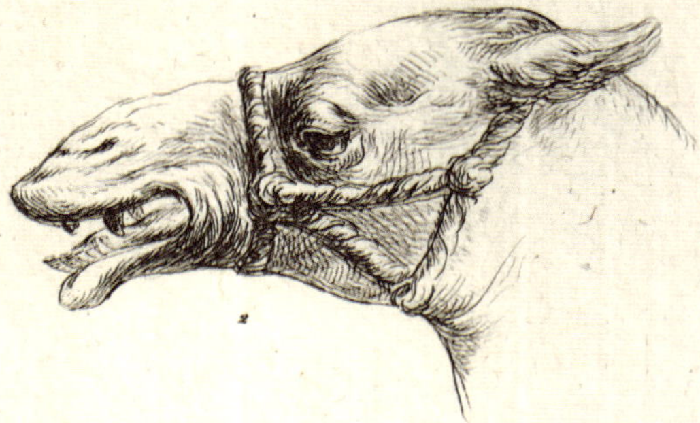

2

Denon Del et Sculp.

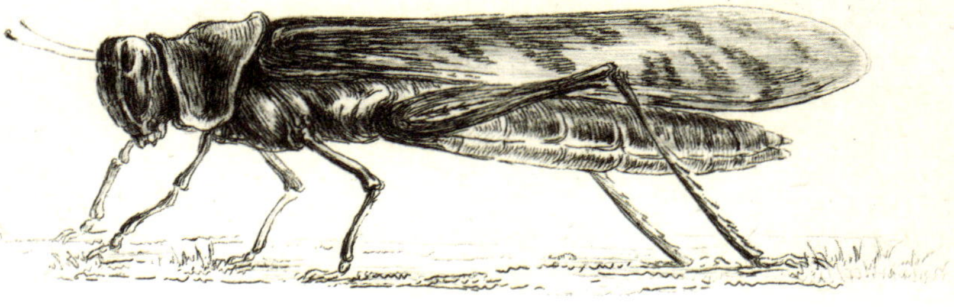

in his descriptions of nature—he mentions the singing of the birds, but very little else—could not have led them to expect much. The ancient monuments were sure to arouse their admiration, but, professionally speaking, how enthusiastic were the naturalists about exploring the natural history of the Upper Nile?

Once on the spot Coquebert, who had not been particularly keen on going south, seems to have thought of the antiquities only[93] and limited his scientific efforts to joining some of his colleagues in inscribing his name on the wall of the temple at Philae. Savigny was, as usual, indefatigable in observing birds. Hypolite Nectoux, who had remained remarkably inactive as a botanist so far, decided to join General Belliard on an excursion by dromedary further south into Nubia to complete his study of the genus *Cassia*, a precious medicinal plant used as a cathartic and emetic whose leaves had been known since the Middle Ages as senna leaves, and which were traded in the markets of Alexandria and Cairo and from there shipped to France. It had soon become clear to him that the dried herbs traded as senna belonged to different species, and in order to decide on their exact identity Nectoux started searching for fresh plants. His search in the vicinity of Alexandria was unsuccessful. Since the presence of wandering Bedouin in the outskirts of Cairo made field work difficult, he sent a native to search for plants in the desert, just as Forsskål had done before him.[94] The stratagem worked. Studying the plants which he thus obtained, he found that yet another species was brought in by the caravans from the south, and decided to set out beyond the cataracts to find where this herb grew. Ultimately, he managed to distinguish the provenance of three different species of Cassia. He also identified the herb which (along with *Cassia lanceolata*) was called Mecca Senna by the local merchants, and whose leaves were used as an adulterant of dried Cassia. They belonged to an unknown plant which Nectoux described in his only publication on Egypt and to which he gave the name 'cynanchum oleae-folium'. The plant, however, corresponded to Delile's *Cynanchum argel*, which, relocated in the genus *Solenostemma*, is today considered its correct name. Arg(h)el, as the plant is known in English, is a common herb in Egyptian apothecary shops sold for coughs, colds, intestinal cramps, inflammation of the urinary tract, and as an antisyphilitic. It has been heavily collected in its natural habitats and has become an endangered species.

It is strange that Nectoux, whose special mission it was to study Egyptian agriculture, did not contribute to the *Description de l'Égypte*—the part on agriculture and commerce is by the engineer Pierre Simon Girard—and that he decided to publish his only essay separately from, but almost contemporarily with, the *Description de l'Égypte*. He himself justifies this by stressing that he wanted to make his discovery known 'to all men' and especially to all physicians and herbalists.[95] The result is a forty-page pamphlet with three fine plates by Henri-Joseph Redouté

Dominique Vivant Denon, snake used by snake-charmers. *Voyage dans la Basse et la Haute Égypte, pendant les campagnes du general Bonaparte*, Paris 1802, plate 104.1
The Arcadian Library, 08962

Dominique Vivant Denon, pufferfish and camel's head. *Voyage dans la Basse et la Haute Égypte, pendant les campagnes du general Bonaparte*, Paris 1802, plate 109.1
The Arcadian Library, 08962

Dominique Vivant Denon, desert locust. *Voyage dans la Basse et la Haute Égypte, pendant les campagnes du general Bonaparte*, Paris 1802, plate 111.1
The Arcadian Library, 08962

NEXT OPENING
Henri-Joseph Redouté, plates of *Cassia lanceolata* and *Cinanchum oleaefolium* (=*Solenostemma argel*). H[ypolite] Nectoux, *Voyage dans la Haute Égypte, au dessus des cataractes : avec des observations sur les diverses espèces de séné qui sont répandues dans la commerce*, Paris 1808, plates 2–3
© Natural History Museum, London, Special Books 581.9 (62) NEC F1

93. '[N]ous avons la tête remplie de chef-d'œuvre de ce peuple singulier.' Letter to his parents, in Doé de Maindreville, quoted in Laissus, *L'Égypte, une aventure savante* p. 300.

94. '[J]e pris le parti d'envoyer un homme du pays me chercher des plantes dans le désert, où l'on ne pouvait pénétrer sans une forte escorte. Forskhael avait employé ce moyen pendant son séjour en Égypte.' Hypolite Nectoux, *Voyage dans la Haute Égypte, au dessus des cataractes : avec des observations sur les diverses espèces de séné qui sont répandues dans la commerce*, Paris 1808, p. 3.

95. Nectoux, *Voyage dans la Haute Égypte*, p. iii ; Nectoux adds : 'Ce résultat de la tâche que je m'étais imposée pendant cette mémorable expédition ; [*sic*] a reçu l'approbation de l'Institut National, dans la séance du 30 thermidor an X, et c'est ce témoignage honorable qui m'a engagé à le publier' (p. iv).

and one by his elder, and more famous, brother Pierre-Joseph Redouté.[96] There is perhaps some irony in the fact that the name he finally gave to the new plant has not been adopted and that the name his more industrious colleague Delile gave it in the *Description de l'Égypte* has been adopted instead.

Henri-Joseph Redouté, plate of the Doum Palm.
Étienne Geoffroy Saint-Hilaire, Jules-César Savigny *et al.*, *Description de l'Égypte*, Paris 1809–12, *Histoire naturelle*, vol. 2 bis, Botanique, plate 1
British Library, 1899.k.1

In his search for new plants in Upper Egypt Delile was quite successful. All in all he collected about 60 species, a dozen of which he believed to be new.[97] The most obvious object for study was the gingerbread tree ('Doum Palm'), which, Denon had reported, was much bigger than the ones they had seen so far.[98] Past descriptions of this tree had largely depended upon its fruits, and not upon its flowers. Delile, however, now had the possibility of seeing its flowers, and found that the Theban or Doum Palm corresponded to the tree Theophrastus had called Cucifera. As Forsskål had felt uncertain about the genus, Delile proposed to adopt Theophrastus' name, and call the palm *Cucifera thebaica*.[99] Delile's research on the Doum Palm is given some prominence in the *Description de l'Égypte*: a separate essay was included in the first volume of the *Histoire naturelle* (1808), and the botanical section of the plates opens with two images of this palm tree, one a fine panorama with Doum Palms, the other a detail of its leaves and fruit. These are executed by Redouté, and they are without doubt the finest plant plates in the book.

The length of Geoffroy's study on crocodiles in the *Description de l'Égypte*—which, with its 40,000 words, is by far the longest essay in the Natural History volumes—seems to indicate that the anatomist's greatest interest in Upper Egypt lay in these huge reptiles. They had been known to inhabit the Nile since the times of Herodotus who had described them and their behaviour in the second volume of his *History*. Denon had seen the first crocodiles near Dandarah. In his *Voyages dans la Basse et la Haute Égypte* he described how, while talking to Desaix, he saw on one of the sandy islets something long and brown amongst numerous ducks.[100] He soon identified it as a sleeping crocodile of about 4.5–5.5 metres. Such measurements sound reasonable. The local population managed to catch a sleeping crocodile which the French wanted to send alive to Paris, partly for study, but above all in homage to the nation's conquest of the Nile.[101] By this time Denon's grandiloquent rhetoric required more imposing specimens, and he says that he had seen animals with a length of almost 8.50 metres. He further reports that some trustworthy officers had seen crocodiles measuring about twelve metres.[102] These fantastic measurements are belied by those of the animal which Redouté drew for the *Description de l'Égypte*, which was only 1.90 metres long, and by those of another bigger specimen, which Geoffroy managed to bring back to the museum in Paris, and which measured no

96. Henri-Joseph Redouté produced the plates of *Cassia sena*, *Cassia lanceolata* and the new species *Cynanchum oleifolium*; the plate of *Cassia fistula* is by Pierre-Joseph Redouté.
97. Amongst those the following are accepted entities today: *Morettia philaeana* (Delile) DC, *Striga hermonthica* (Delile) Benth., *Crotalaria thebaica* (Delile) DC, *Psoralea plicata* Delile, *Acacia albida* Delile, *Tephrosia apollinea* (Delile) Link. The specimens of *Morettia* and *Acacia* were collected by Nectoux and described by Delile.
98. Denon, *Voyages dans la Basse et la Haute Égypte*, I.141.
99. Delile's genus has been rejected recently, and the Theban Palm is called *Hyphaene thebaica* (L.) Mart. today.
100. Denon, *Voyages dans la Basse et la Haute Égypte*, I.140.
101. '[D]'en faire un hommage à la nation comme un trophée de la conquête du Nil.' Denon, *Voyages dans la Basse et la Haute Égypte*, I.204.
102. Denon, *Voyages dans la Basse et la Haute Égypte*, I.209.

more than 2.86 metres.[103] Although Alpini, Hasselquist and Norden all speak of crocodiles about ten metres long, it is clear that Geoffroy did not see one.[104]

Geoffroy opens his essay on the crocodile with Herodotus. He quotes the entire passage from *Euterpe* and then proceeds to comment on the historian's description line by line. The most interesting part is the comment on the *trochilus*. Herodotus writes:

Since it lives in the water, its mouth is all full within of leeches. All birds and beasts flee from it, except only the sandpiper [trochilus], *with which it is at peace, because this bird does the crocodile a service; for whenever the crocodile comes ashore out of the water and then opens its mouth (and this it does for the most part to catch the west wind), the sandpiper goes into its mouth and eats the leeches; the crocodile is pleased by this service and does the sandpiper no harm.*[105]

This passage has puzzled many commentators, Geoffroy dryly comments. Some have seen it merely as a product of the imagination, others have tried to vindicate the author by arguing that the *trochilus* can inflict its will on the crocodile,[106] while Aristotle, in his *History of Animals*, maintained that the so-called *trochilus* actually cleaned the reptile's teeth.[107]

As no one had as yet identified Herodotus' *trochilus*, Geoffroy hoped that his excursion into Upper Egypt would shed some light on this mysterious bird. He soon found it corresponded to the bird Hasselquist had described as *Charadrius Aegyptius*,[108] a small insect-eating plover which could not possibly swallow a leech. Moreover, leeches did not occur in the moving waters of the Nile. Observing a dead, but still fresh, animal, Geoffroy found that the bird did not free the crocodile from leeches nor did it clean its teeth.[109] It simply fed on the insects that were attracted by the brownish crust in the crocodile's oral cavity, and being nearly tongueless, the huge reptile let the small bird do what it could not do itself: chase away the insects. He expresses this basically evolutionary concept of symbiosis as follows: 'Nature had thus created crocodiles without the means to look after their own welfare, or their conservation [...] Therefore, in the absence of a complete organization, nature came to the help of the crocodile.'[110]

In the rest of his long essay on crocodiles Geoffroy distinguishes five different species for the Nile, but admits that he had found the classification of crocodile

103. *Description, Histoire naturelle*, I.259.
104. Compare *Description, Histoire naturelle*, I.192.
105. Herodotus, *History*, II.68, p. 357.
106. *Description, Histoire naturelle*, I.198–9.
107. Aristotle, *The History of Animals*, IX.6.
108. = *Pluvianus aegyptius* (Linnaeus, 1758).
109. The ornithologists Richford and Mead (2003) endorse Aristotle's version, namely that the Egyptian Plover removes little pieces of meat from between the crocodile's teeth (Andrew S. Richford, and Christopher J. Mead, 'Pratincoles and Coursers', in Christopher Perrins, ed., *Firefly Encyclopedia of Birds*, Buffalo 2003, pp. 252–3).
110. 'La nature auroit donc établis les crocodriles sans les moyens de pourvoir personnellement à leur bien-être, aux soins de leur conservation [...] Ainsi, à defaut d'une organisation complète, la nature seroit venue au secour de crocodile.' *Description, Histoire naturelle*, I.203–4.

species the most difficult task in his zoological career.[111] Besides Cuvier's *Crocodylus vulgaris*, he found that two other species existed, while two more species from ancient mummified individuals could be described. None of these seem to have any taxonomic validity and are now all grouped under *C. niloticus* Laurenti, although some zoologists accept Geoffroy's *C. suchus* as a subspecies of *C. niloticus*. As with Nectoux's finds in Upper Egypt, it is ironic that Geoffroy's longest, and most difficult, contribution to the *Description de l'Égypte* should have remained ultimately unsuccessful in terms of taxonomy. What his study of crocodiles would do, however, was to strengthen his theory of homologies later in his career.

Geoffroy did not study the crocodiles in detail on the spot, and much of his anatomical work was done on specimens preserved in alcohol when he had returned to Paris. Although he continued collecting, his concentration seemed to have diminished. The main reason for this is that the expedition in Upper Egypt suddenly came to an end. On 16 September news reached the savants that Bonaparte had left with Monge, Berthollet and Denon for France. Disenchanted with the idea of further research, they all decided to return north the very same day. Contrary winds made sailing down the Nile slow and they only reached Cairo on 6 November. As the English fully commanded the Mediterranean, prospects looked bleak. Although their patron Bonaparte had eluded the English navy and had reached Paris in safety, they found themselves in exile in Egypt. The army looked with disdain and ridicule on the returning savants, blaming them for having been the prime reason for the whole expedition. All that remained was to seek a disadvantageous peace treaty with the English and the Turks. And to know that their lot (and the fruits of their work) were entirely in the hands of the enemy was not comforting. Geoffroy wrote in a 'longue jérémiade' to Cuvier that it was frightful for a French heart to know what had become of an expedition which had set out with the fairest prospects.[112]

Suez and the Red Sea

Back in Cairo, most of the savants felt that their task of collecting information about Egypt had been fulfilled and they expected to be sent back to France, as Bonaparte had instructed Kléber to do. This was not to be. The new general-in-chief was in no position to effect such a return journey right then. Instead, having seen the wealth of material that had been accumulated so far, he ordered it to be condensed in one work.[113] The idea for what was to become the monumental *Description de l'Égypte* was not all his: on 12 October Fourier had written to him about 'a complete work on Egypt'.[114] The project was received with enthusiasm, and Geoffroy prophesied that 'the time will come when the work of the committee of arts will, in the eyes

111. 'La détermination de ce crocodile et des espèces suivantes fut, dans ma carrière zoologique, ce que j'ai encore trouvé de plus difficile à faire.' *Description, Histoire naturelle*, I.258.

112. '[I]l est affreux pour un cœur françois de connoitre l'issüe d'une expédition faite sous d'aussi heureux auspices, de prévoir qu'une des plus belles armées de la république aura été en partie détruite sans que la république en ait tiré d'autre fruit que d'avoir enhardi ses ennemis!' Geoffroy to Cuvier, 27 September 1799, Geoffroy, *Lettres écrites d'Égypte* pp. 144–5.

113. '[U]n ouvrage commun.' *Décade*, III.300.

114. '[U]n ouvrage complet sur l'Égypte.' Quoted in Laissus, *L'Égypte, une aventure savante*, p. 299.

of posterity, excuse the frivolity with which our nation, so to speak, rushed into the Orient.'[115] This was some consolation for the disdain and ridicule with which the savants were greeted when they returned to Cairo from Upper Egypt. So during the rest of 1799 and the beginning of 1800 many members of the committee started to work on the description of modern Egypt. At this stage, the naturalists do not seem to have been included in the great work on Egypt and were expected to continue to amass specimens. For this purpose they went to Suez towards the end of the year. The prospect of exploring the Red Sea no longer delighted Geoffroy. He was convinced he would never see his parents or his friends in France again. It made him physically ill. Feeling truly abject he wrote on 22 December to Cuvier that Egypt had become intolerable to him.[116] Over the past year he had had numerous specimens from the Red Sea sent over to him in Cairo and he did not expect to obtain any new species. His only reason for going to Suez was the fear of later being accused of having neglected an important opportunity to complete his research.

On the way to Suez some final collecting was done. Delile explored the desert in all directions as well as the coastal areas of the Red Sea. Indeed, a considerable number of the plants the expedition collected in Egypt were found during these weeks. About a sixth of the 150 species described in 'Flore D'Égypte'—which is the section where Delile provided elaborate descriptions of new species, new combinations or species that needed further confirmation—refer to plants collected during the excursion to Suez. The majority of these were described as new. They include a number of entries that are accepted today, as, for example, the shrub *Cleome droserifolia* (Forssk.) Delile, the tree *Acacia seyal*, and the grass *Dichanthium foveolatum* (Delile) Roberty. His description of *Cornulaca monacantha* constituted a new genus.[117]

Savigny too was at great pains to procure a good collection of invertebrate sea beasts. The Gulf of Suez and the Red Sea provided him with 15 species of red-blooded worms, many of which were to constitute new genera when he worked out his system of annelids years later. The rich collections of molluscs, madrepores and sea urchins he made in Suez were the basis of the many plates that he prepared with care for the second volume of illustrations of the *Histoire naturelle*, but which, unfortunately he would never be able to annotate.

Geoffroy Saint-Hilaire employed four fishermen to get him the rarest species of fish. 'No work is left undone,' he assured the professors at the museum, but added, 'Here I conclude all my research.'[118] He had received news that the English had agreed to allow a neutral ship escort the naturalists (and the material they had amassed) back to France.

115. '[I]l arrivera que l'ouvrage de la commission des arts excusera aux yeux de la postérité la légèreté avec laquelle notre nation s'est, pour ainsi dire, précipitée en Orient.' Geoffroy to Cuvier, 27 September 1799, Geoffroy, *Lettres écrites d'Égypte*, p. 147.

116. 'L'Égypte m'est insupportable.' Geoffroy, *Lettres écrites d'Égypte*, p. 152.

117. Delile acknowledges he had also collected the species in Upper Egypt, and later in 1801 also found it around Saqqarah (*Description, Histoire naturelle*, II.207).

118. 'Rien n'est négligé ... Je termine ici toutes mes recherches.' Geoffroy to his colleagues at the Muséum, 7 January 1800, Geoffroy, *Lettres écrites d'Égypte*, p. 156.

Alexandria and Cairo

Notwithstanding the news they received at Suez, the Egyptian adventure did not end so smoothly. First there was confusion about whether Cairo was still in the hands of the French and when the savants reached Rosetta the plague had visited the town and they were quarantined for over a month. When they finally got to Alexandria in the middle of March, negotiations with the English for their passage had stalled. Their passports had been cancelled and they could not embark for France. Waiting in vain at Alexandria the naturalists tried to make the best of their time, collecting specimens in the area around the town. Delile and Savigny found many desert species of plants and insects, while Geoffroy dissected the salt-water fish of the Mediterranean. But there was no spirit in it all and they felt irascible. He wrote that the entrails of fish could not hold his interest and he often thought with regret of 'those good days of my life when we amassed new facts and lived without noticing the bustle of society and the petty emotions of man'.[119]

Redouté too tried to make the best of his time. In the library of the Muséum National d'Histoire Naturelle in Paris there are four water-colour drawings of salt-water fish executed by Redouté at Alexandria which were used for plates in the *Description*.[120]

In June the savants were ordered back to Cairo, where Geoffroy felt they were completely useless. They had seen everything they wanted to see in Egypt, and were looking forward to start organizing their collections. But as they did not have a good library at their disposal, this could only be done in Paris. Accordingly all the collections were packed, and the naturalists had to wait for new orders.

The savants also suffered from ill health. Geoffroy had recurrent attacks of dysentery, and ophthalmia rendered him nearly blind for almost a month. Delile too was often ill, and Nectoux had lost nearly all his eyesight. Notwithstanding ill health and disillusionment, the sessions at the Institut were resumed and a great variety of papers were read. At the beginning of 1801 Geoffroy participated in an exploration of the pyramids of Saqqarah. These had not been penetrated as yet, and for the first time the French saw a pyramid intact with the mummified remains of humans and of birds, as well as many curious artefacts and articles of clothing. Jacques Menou, who had become general-in-chief after the assassination of Kléber, spared no means or money to excavate the site. Geoffroy drew up an instruction in the general's name that an inventory be made of all manuscripts, mummies, tools, vegetal remains and artefacts found.[121] It was only on 6 April 1801 that the members of the Committee could prepare to move to Alexandria again. Coquebert was not amongst them. He died the following day of the plague.

As in the previous year, the collections were immediately stowed in the hold of a ship upon entering Alexandria, but then there was further delay. There was nothing

119. 'Ce n'est pas que souvent les viscères des poissons ne m'offrent des faits très intéressans et alors je suis replacé à cette belle époque de ma vie, où nous nous occupions ensemble d'amasser de nouveaux faits et vivions sans nous apercevoir des tracasseries de la société et de petites passions des hommes.' Geoffroy to Cuvier, 30 April 1800, Geoffroy, *Lettres écrites d'Égypte*, p. 179.

120. *Description, Histoire naturelle*, vol. I, plates 22.1, 22.2, 25.1–2, 25.3–4.

121. Geoffroy, 'Rapport a l'Institut d'Égypte sur les recherhes à faire dans l'emplacement de l'ancienne Memphis, et dans toute l'étendue de ses sépultures', in Geoffroy, *Lettres écrites d'Égypte*, p. 252.

more exasperating than this wait. Geoffroy's skins of birds and stuffed animals were ruined, and Savigny's collections of insects suffered greatly. Moreover, the spirits that were used to preserve the specimens were slowly evaporating. 'If my stay is prolonged a little longer,' he wrote to Cuvier, 'I will only bring you mummies.'[122]

In 1801 the English invaded Egypt. After the disastrous battle of Canopus and the capitulation of the French army in Cairo, Menou was forced to retire to their last stronghold, Alexandria. Besieged by English forces both on land and at sea, the French army surrendered in August. But it was only in October 1801 that the naturalists could board ship for France. With Menou's surrender General Hutchinson claimed the collections of the members of the Committee for England. Menou, who wanted to keep the Rosetta Stone, did not mind forfeiting the collections of the naturalists. Savigny and Delile protested, and Geoffroy, interrupted in his research on electric fish, declared he would not give up his collections. Rather he would follow them to England, or even destroy them completely. The general wrote ungracefully to Hutchinson: 'I have just heard that many of the collectors desire to follow their seeds, minerals, birds, butterflies or reptiles, no matter where you send the crates. I do not know if, in order to do so, they will have themselves stuffed, but I can assure you, that, if the fancy takes them, I will not stop them from doing so.'[123] However, Hutchinson listened to the naturalists' arguments for keeping their collections, and agreed that, except for the Rosetta Stone, their collections could proceed to France for further study. The naturalists reached France on 18 November 1801.

122. 'Pour peu que mon séjour se prolonge, je ne vous rapporterai que des momies.' Geoffroy to Cuvier, 20 September 1801, Geoffroy, *Lettres écrites d'Égypte*, p. 203.

123. Quoted in Laissus, *L'Égypte, une aventure savante*, pp. 397–8.

CHAPTER 3

EHRENBERG AND HEMPRICH: THE PRUSSIAN EXPEDITION OF 1820–1825

Berlin

In 1820 the art-collector and Egyptologist Johann Heinrich Karl Menu von Minutoli started on an antiquarian expedition to Egypt. After a successful military career in the Prussian army and as captain of the Berlin Cadettencorps, Minutoli had been appointed tutor to Prince Karl, in whom he instilled a passion for antiquities. His collections as well as his knowledge of archaeology and ancient history earned him a reputation in the field, and when he decided on an expedition to the East, King Friedrich Wilhelm III showed interest. This was the time of the political and cultural reform of the Prussian state after its collapse following the defeat at the hands of Napoleon in 1806, and, more recently, the repartition of its traditional territories by the Congress of Vienna (1814–15). Part of this reform was rooted in a new nationalistic awareness, which was both a reaction against Napoleonic dominance and the consequence of the Romantic interest in history, and which was voiced, amongst other things, in a strong emphasis on education. Thus Friedrich Wilhelm founded in 1809 the University of Berlin, in 1811 the University of Breslau, and in 1818 the University of Bonn. It was especially the Friedrich-Wilhelms-Universität in Berlin (today the Humboldt University) which soon enjoyed an almost matchless reputation in Europe.

Minutoli's enterprise was a private undertaking, but the King of Prussia fully supported the plans. Although a trip of this kind would furnish precious strategic information about the growing economic potential of the Near East for Prussia, it was in the cultural framework of Prussian academic excellence that Minutoli's expedition to Egypt and the Near East must be seen. By buying up the collections, the University of Berlin would have an opportunity to found its own Egyptian museum and, in the European context of learning, contribute significantly to the study of Ancient Egypt. From such a point of view it is not surprising either that the Prussian government wanted to expand the scope of the expedition beyond archaeology and add a small group of scholars and scientists. Upon Minutoli's suggestion the Royal Academy of Sciences decided to finance two naturalists who would be attached to the expedition, and the choice fell on the 24-year-old zoologist Wilhelm Friedrich Hemprich and the 25-year-old botanist Christian Gottfried Ehrenberg. Moreover, the Ministry of Culture and Education allowed Professor Ludwig Theodor Liman, an architect, and August Scholz, a philologist from the University of Bonn, to travel with Minutoli.

Ehrenberg was a promising young naturalist from Delizsch. After graduating in theology at the University of Leipzig, he devoted himself to the study of medicine and natural science. In 1817 he was drawn to the Prussian University of Berlin, where, in 1818, he presented a work on fungi (*Sylvae mycologicae berolinensis*) in which he described numerous new species. He befriended the renowned naturalist Alexander von Humboldt and, influenced by the explorer's expedition to South America, dreamt of travelling the world himself. It was no good being a closet naturalist. 'The idea of a voyage is my ideal,' he wrote in 1819 to the botanist Christian Gottfried Nees von Esenbeck, and added: 'It is my firm belief that all good for mankind must derive from natural history.'[1]

Hemprich was another product of Prussia's Enlightenment project. He had studied medicine first at the University of Breslau and later at the University of Berlin. He was asked to lecture on comparative physiology, a subject on which he published in 1820 the treatise *Grundriss der Naturgeschichte*. He was much interested in the amphibians and reptiles at the zoological museum, which he studied under its director Hinrich Lichtenstein. In Berlin he also met Ehrenberg, and the two naturalists became friends. Hemprich too had long desired to travel and the two planned an expedition to Madagascar, but nothing came of it. When he heard of Ehrenberg's appointment as naturalist on Minutoli's expedition, Hemprich lost no time in pleading with Lichtenstein for funds to join the expedition in the name of the museum. Lichtenstein set to work, and on 23 March 1820 the Royal Academy of Sciences decided to sponsor both Ehrenberg and Hemprich. On 7 June it drafted an 'Allgemeine Instruction' in which the travellers were told to attach themselves to Minutoli's expedition. An ominous clause was inserted that they should in no way be a burden to the general and always pay him due respect. Even so, although being escorted by Minutoli had obvious advantages, the naturalists were not bound to stay with him if they deemed excursions to other areas necessary. The Academy further advised them not to visit those countries, however interesting, where communication with Prussia was uncertain: conveying their collections back to Berlin at regular intervals was seen as a priority. They were also asked to forward their diaries and notes. The Academy would make sure that important discoveries would be announced to the world of science, but that all the details would be safeguarded till the two naturalists could work them out themselves after their return. For the expenses the Academy granted them the modest sum of 3,500 Reichstaler.

To ensure the success of the expedition, the Prussian Prime Minister Karl August von Hardenberg decreed that the two naturalists could count on the special collaboration of all the Prussian consulates in the countries they were to visit, and in Paris Alexander von Humboldt procured them a series of letters of introduction to many persons of influence in the towns in the Near East. When Ehrenberg and Hemprich left Berlin on 15 June 1820, and travelled via Breslau to Vienna, they sought out the naturalists at the universities there and received further advice for research in the Levant. In Vienna they even managed to obtain from the minister of foreign affairs Metternich a letter of introduction to the Austrian consulate in Egypt.

Everything seemed well organized. When Ehrenberg and Hemprich reached Trieste on 2 August and met Minutoli, however, their problems started. Although it had been agreed they would all sail with Minutoli on 15 August, leaving them

1. 'Die Idee einer Reise ist mein Ideal. Ich denke nun einmal, daß alles Heil der Menschheit aus der Naturgeschichte entspringen muß.' Ehrenberg to Nees von Esenbeck, 5 September 1819, Erwin Stresemann, *Hemprich und Ehrenberg, Reisen zweier naturforschender Freunde im Orient geschildert in ihren Briefen aus den Jahren 1819–1826*, Berlin 1954, p. 11.

plenty of time for their final preparations, they now found to their surprise that they only had three days to make preparations before sailing. What had happened was that the general had quite unexpectedly decided to get married in Trieste and take his bride and entourage with him to the East. This meant that there was no longer any room on the ship for the naturalists, and as no other ships would sail for Alexandria for some time to come they could only start with *Il Filosofo* on 5 August. As a result, Hemprich complained to Lichtenstein that all their purchases were made in haste, that many indispensable items were forgotten, and that, as they now had to arrange for a ship themselves, a considerable part of their funds had been spent.[2] It would soon become clear that Minutoli was not a suitable and reliable patron for their voyage to the East.

Alexandria

The *Filosofo* sailed south along the islands of the east Adriatic. As there was no wind they anchored for days in Kotor (Montenegro), and while Hemprich had a respite from the seasickness from which he had been suffering, Ehrenberg sailed up the gulf in search of plants. Although the stark mountains around the bay had little to offer at this time of year, he came up with a detailed description of 'the botanical character of the area in the month of August', listing 35 cultivated and wild species, including a couple of unidentified savoury plants which he thought worth describing as new.[3] More impressive were the 166 species of insects he managed to collect during the short stay in the area, many of them known southern European species, but some of them as yet undescribed.[4] Ehrenberg's account of Kotor ends with a long description of its people and their habits. The delay in Montenegro was hard to bear, however, and Ehrenberg felt relieved when, on 23 August, the long-awaited day of departure came.[5]

On 4 September they viewed Pompey's column and soon after they entered the harbour of Alexandria. Ehrenberg's first impressions were the 'black and brown, mostly naked, Arabs' in the harbour and the 'completely naked or ragged, and very dirty, children' in the town.[6] As the Prussian consul was absent from Alexandria, Hemprich set out with Metternich's letter of introduction to the house of the Austrian consul, where the naturalists were invited to stay until the arrival of Minutoli. As soon as they had moved their belongings to the Austrian consulate, they started to explore the flora and fauna around Pompey's column. The beginning of their research on the African continent was not very promising. Not a single plant could be found, and they only managed to shoot four birds. The sixty insects and

2. Hemprich to Lichtenstein, 1 October 1820, Stresemann, *Hemprich und Ehrenberg*, p. 17.
3. The new species were *Satureja campanella* and *S. olla* (see Christian Gottfried Ehrenberg, *Naturgeschichtliche Reisen durch Nord-Afrika und West-Asien 1820 bis 1825 von W. F. Hemprich und C. G. Ehrenberg*, Berlin 1828, p. 13). Neither has any validity today.
4. Unlike the new *Satureja* species Ehrenberg does not attempt any description of the new insect species in *Naturgeschichtliche Reisen* (see p. 15).
5. Ehrenberg, *Naturgeschichtliche Reisen*, p. 19.
6. '[S]chwarzer und brauner, meist nackter Araber'; 'Ganz nackte oder mit Lumpen behangene, höchst schmutzige Kinder'. Ehrenberg, *Naturgeschichtliche Reisen*, pp. 25–6.

ten lizards they caught were a meagre consolation: 'we knew that we had come to a strange, albeit barren, nature.'[7] The following day was more successful, but as they came back from the new harbour they attracted the attention of the local people: 'We came back in our European summer clothes loaded to our hearts' content with insects, amphibians and salt-water plants, and a group of loudly screaming naked Arab youngsters accompanied us.'[8] Their enthusiasm diminished when they returned to the Austrian consulate to find that ants had completely ravaged their insect collection overnight. The very same day Minutoli arrived. But before moving on with the general's troupe there was still plenty of time during the following days to explore the Cape of Figs. Forsskål's descriptions had led them to expect much, and they were not disappointed. They came away with their botanical tins packed with plants.

The first part of Minutoli's expedition was to visit 'famed Cyrenaica', the ancient Roman province of east Libya, and from there to travel to Cairo. To this proposal the two naturalists enthusiastically agreed, and they already had visions of collecting 'the golden fruits of the gardens of the Hesperides'.[9] At least this is what the published account of the expedition in *Naturgeschichtliche Reisen* says. Ehrenberg added that, as Minutoli's interest in archaeology and antiquities often meant stopping in a place for some time while the naturalists needed to cover as much distance as possible to collect specimens, it was unacceptable for them to follow Minutoli's caravan. This account, however, concealed the growing disagreement between the naturalists and Minutoli. According to the original plan they were to travel directly by boat to Cairo, which could be arranged cheaply with the help of the local Pasha. Minutoli, however, suddenly changed his mind and proposed organizing a caravan for Cyrenaic Darnah, and travel from there to Cairo. He assured the naturalists that the detour would cost little more than the boat to Cairo, and, attracted by the possibility of exploring the Libyan desert, they agreed. They ordered camels, bought Bedouin dress, arms, and waterbags, and fitted their camels out with wooden crates to carry their collection of bottles with spirits. They also engaged an interpreter. The total sum they spent fell well within their budget, but they had not reckoned with Minutoli's travelling standards. Hemprich realized that Minutoli wanted to use them 'to travel cheaply, yet *à la General*'.[10] What had happened was that the general had decided to travel with 50 servants and 32 camels and that he expected the naturalists to pay their share of the expenses. Hemprich had to tell him that they would not pay more than 1,000 florins[11] and remind him that their Instruction stipulated that they were expected to follow the general only as long as this was reasonably possible. 'As for the General,' Hemprich lamented to Lichtenstein, 'notwithstanding his possibly

7. '[...] zwar in einer fremden, aber in einer kargen Natur befanden.' Ehrenberg, *Naturgeschichtliche Reisen*, p. 29.

8. 'Wir waren, in europäischer Sommerkleidung, bei unserer Rückkehr mit Insecten, Amphibien und Seepflanzen zur Zufriedenheit beladen, und es begleitete uns ein ganzer Trupp nackter arabischer Jugend mit hellem Geschrei.' Ehrenberg, *Naturgeschichtliche Reisen*, p. 32.

9. Ehrenberg, *Naturgeschichtliche Reisen*, p. 44.

10. 'um selbst wohlfeiler und doch *à la General* zu reisen'. Hemprich to Lichtenstein, 23 September 1820, Stresemann, *Hemprich und Ehrenberg*, p. 16.

11. They had expected to pay about 600 florins, while a later fourteen-day excursion all by themselves cost them only 150 florins (Hemprich to Lichtenstein, 1 October 1820, Stresemann, *Hemprich und Ehrenberg*, p. 17).

good intentions, he has done little to further our journey, much to damage it.'[12] All the same, the first 5 days in the desert were successful: Hemprich reported that they collected 400 insects, 13 reptiles (including three unknown species), seven birds, and a great number of interesting plants and lichens. But before these could be studied, the collection was packed in crates and sent from Alexandria to Berlin.

Minutoli had organized the detour to Darnah badly, and when the Arabs accompanying them lost time in getting to the frontier, where they waited in vain for the Bey of Darnah's permission to travel onward, the General decided to forgo the Cyrenaican plan and deviate to Siwa Oasis, 350 miles south-west of Cairo. As travelling through the desert in autumn was not attractive to the naturalists, they decided at this point to leave Minutoli's caravan and continue to Darnah on their own. Scholz and Liman too preferred to await the messenger from Darnah and then proceed to Cyrenaica. Three weeks elapsed before word came from the Bey. Afraid as he was of spies, he let them know that he wanted advice from Tripoli first, which could not be expected for another 50 to 60 days. As it was too dangerous to stay for long on the frontier, the small group had to follow in Minutoli's footsteps. Ehrenberg and Hemprich felt disappointed that they would not see Cyrenaica, but the oasis of Siwa had never been studied by naturalists and was promising. However, they were disappointed again: Minutoli, who was there before them, had managed to rouse the natives's hostile feelings against all Christians. His indiscreet way of making drawings had created so much unrest that violence was feared and he had to flee the area.[13] As a result the Prussian naturalists were not welcomed in Siwa, and it was with bitterness that they realized that no collecting was to be done in the oasis and that they had better return to Alexandria. By this time Liman and Wilhelm Söllner, Ehrenberg's friend and relative who had volunteered to join the expedition, had fallen so ill that they had to be bound on their camels, and Hemprich suffered from an incapacitating toothache. The camels were recalcitrant and threw off their loads of specimens at least three times a day, breaking half the contents. Not a day passed without loss, and when they finally reached Alexandria on 9 December in pouring rain after a journey of great hardship, morale was low. Liman died a couple of days later, and was buried in the Greek cemetery. It was in such a mood that they received a bill from Minutoli in Cairo detailing their expenses for the detour to Darnah, which he expected them to pay forthwith. To Hemprich it was now clear that the General, who had been too proud to accept financial support form the Prussian government for his expedition, was after the scholars' money.[14] It also appeared that Minutoli had sequestrated and transferred their funds. According to Ehrenberg's published account this was because he expected them to travel from Darnah to Cairo.[15] Hemprich was apparently of a different opinion. Whatever the correct explanation, they found themselves in Alexandria without any money.

Before moving to Cairo, Ehrenberg and Hemprich prepared a crate with their

12. 'Was den Herrn General betrifft, so hat er uns bey unserer Reise, trotz vielleicht vorhandenem guten Willen, nichts genützt, viel geschadet.' Hemprich to Lichtenstein, 1 October 1820, Stresemann, *Hemprich und Ehrenberg*, p. 17.

13. In *Naturgeschichtliche Reisen* Ehrenberg only blames the absence of the Ottoman governor of Siwa for the hostilities against Christians (pp. 120–1).

14. Hemprich to Lichtenstein, 23 December 1820, Stresemann, *Hemprich und Ehrenberg*, pp. 23–4.

15. Ehrenberg, *Naturgeschichtliche Reisen*, p. 148.

Hand-coloured lithograph of *Truxalis pharaonis* (= *Truxalis annulata* Thunberg 181–5). Christian Gottfried Ehrenberg, *Symbolae physicae seu Icones et descriptiones corporum naturalium novorum aut minus cognitorum quae ex itineribus Libyam Aegyptum Nubiam Dongalam Syriam Arabiam et Habessiniam ...*, Pars Zoologica II, Insecta, Berlin 1830, plate XIV
Linnean Society Library, FF 916.1/.3:57 EHR

collections of animals to be forwarded to Berlin. It contained five mammals, 48 birds, 123 reptiles, seven fishes, 300 shells, 500 worms, and 180 minerals. Notwithstanding these elevated numbers of specimens, there were only a few new species, which were described and published by Lichtenstein in 1823.[16] Amongst these were a few lizards, one bird (the wheatear *Oenanthe moesta*) and two rodents: the Libyan Jird (*Meriones libycus*), and the Four-toed Jerboa (*Allactaga tetradactyla*). This last species has been placed on the *IUCN Red List of Threatened Species*.[17] Its specific habitats along the coast of Egypt and Libya have been degraded and the species is believed to be in decline.

On 20 December rumours of the plague reached the naturalists and a few days later they were taken into quarantine in the Prussian consulate. It was only on 23 January that they could finally embark for Cairo.

Ehrenberg used the first opportunity to go on land in the Nile Delta. From the descriptions of travellers who had gone before them he had high expectations, but he returned feeling tired and gloomy from his first excursion. There were no crocodiles, but only a handful of European plant species, and a few European bird and insect species. 'I reluctantly threw down the German nettle, which burnt me in Egypt,' he grumbled in his travel account.[18] And he concluded: 'The groups of palm trees and the shapes of the naked blackish-brown Arabs can interest the European for some time, but the characteristics of the vegetation are not as unexpected as to engross the mind.'[19] Shortly after reaching Cairo on 4 February 1821, Ehrenberg's relative Söllner died. The toll of the first part of the Egyptian expedition had been high, and Ehrenberg and Hemprich had understandably very mixed feelings about the success of their scientific mission so far.

Cairo, Saqqarah and the Fayyum

In February the spring season was starting in Cairo, and Ehrenberg reported to Professor Link that they therefore concentrated mainly on collecting plants. They managed to assemble a rich collection, but without any remarkable new forms.[20] Despite their work on the flora of Cairo, they also managed to obtain specimens of two-thirds of all known species of Nile fish (including three specimens of Geoffroy's Nile Bichir) and lots of reptiles.[21] Hemprich reported a large unknown weasel, which Ehrenberg included in *Symbolae physicae* in 1833 as a subspecies of the

16. M.H.C. Lichtenstein, *Verzeichnis der Doubletten des zoologischen Museums der Königliche Universität zu Berlin nebst Beschreibung vieler bisher unbekannter Arten von Säugethieren, Vögeln, Amphibien und Fischen*, Berlin 1823.

17. Aulagnier, S. 2004. *Allactaga tetradactyla*. In: IUCN 2006. *2006 IUCN Red List of Threatened Species*. <www.iucnredlist.org>. Downloaded on 14 December 2008.

18. 'Unwillig warf ich die deutsche Nessel zu Boden, die mich in Aegypten brannte.' Ehrenberg, *Naturgeschichtliche Reisen*, pp. 153–4.

19. 'Die Palmengruppen und die Formen der nackten schwarzbraunen Araber können wohl die Phantasie eines Europäers eine Zeit lang beschäftigen, aber die Eigenthümlichkeit der Vegetation ist zu wenig überraschend, als daß sie begeistern könnte.' Ehrenberg, *Naturgeschichtliche Reisen*, p. 155.

20. 'ohne jedoch auf sehr auffallende neue Formen gestoßen zu seyn.' Ehrenberg to Link, 25 March 1821, Stresemann, *Hemprich und Ehrenberg*, p. 30.

21. Hemprich to Lichtenstein, 7 March 1821, Stresemann, *Hemprich und Ehrenberg*, p. 27.

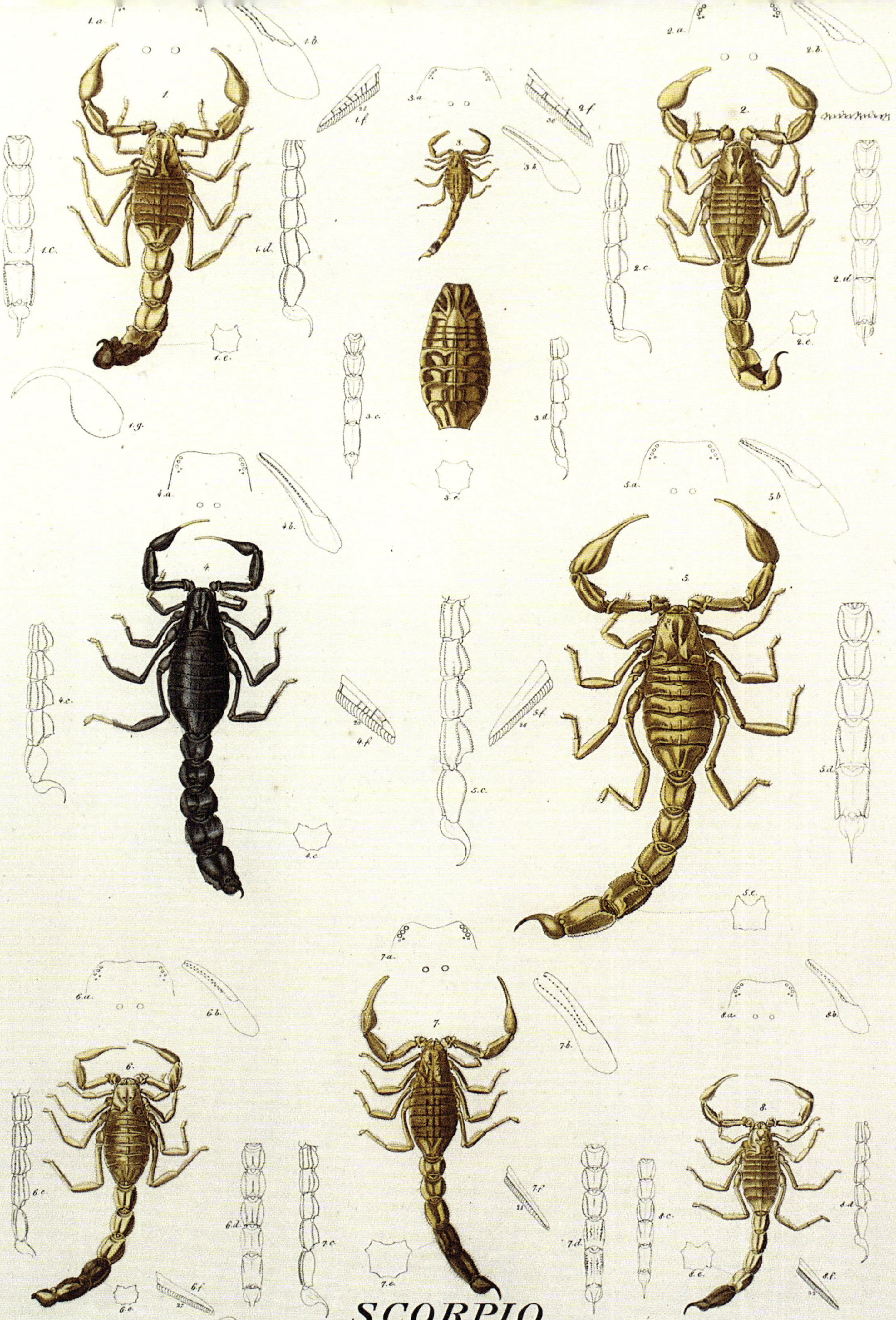

SCORPIO

1. ANDROCTONUS (Prionurus) *libycus.*	5. ANDROCTONUS (Prionurus) *funestus.*
2. . . . . *citrinus.*	6. . . . . *leiosoma.*
3. . . . . *nigrocinctus.*	7. . . . . *scaber.*
4. . . . . *bicolor.*	8. . . . . *melanophyssa.*

Sam. Weber ad. Nat. del. et sculp.

Common Weasel (*Mustela nivalis subpalmata*) but which some authors today regard as a separate species.[22]

On 2 April they went south to the Fayyum. Ehrenberg, however, fell severely ill, and they had to set up camp at Saqqarah, waiting for him to convalesce. While week after week passed without any sign of improvement in his condition, Hemprich collected insects and birds. On 23 April he boasted about 800 insects and 25 birds—adding that, if Ehrenberg had not fallen ill, they would certainly have doubled this.[23] By 30 April he had 1,200 insects and 40 birds. Although he felt elated with the much-cherished specimens of species of hyena, bat and vulture, Ehrenberg's inability to recover cast a gloom over the results. His situation was truly unpleasant, he wrote to Lichtenstein on 5 May: he had to do all the work alone, hemmed in by the plague, and with no news from Berlin. And he added in a postscript, 'Were I not to return, you will put in a word for my mother, won't you?'[24] When Hemprich was struck with ophthalmia, they decided to abandon the Fayyum project for the time being and return to Cairo.

Ever since his letter to Lichtenstein of 1 October 1820 Hemprich had made it clear that their financial resources were utterly insufficient, and he stressed the necessity of further sums of money. On 25 February Lichtenstein wrote that the Academy had no funds and that none could be obtained from the government to finance their expedition. Although the Prussian minister of education Altenstein had been entreated by the professor of anatomy Karl Asmund Rudolphi for financial support, only what little money could be spared at the Museum was forthcoming at that stage. This disappointing letter reached Ehrenberg and Hemprich towards the end of May 1821, when Hemprich was about to try for the Fayyum a second time, leaving Ehrenberg to recover in Cairo. The spectre of finding themselves in Egypt without enough money to return to their country made it necessary to be as explicit as possible. If no money reached them by the end of October, Hemprich answered Lichtenstein, they would take this as proof that the people in Berlin did not appreciate their research any longer and wanted them to return to Germany. It would mean the complete failure of the Prussian expedition, he concluded: 'Professor Liman is dead, the General is coming back rather empty-handed, and when we too go home, then the whole expedition = null.'[25] This letter had an effect. The enthusiasm for the collections they had forwarded to Berlin assured them new credit, although they would not know this until the summer of 1822.

The second attempt to naturalize around the Fayyum was hardly more successful than the first. On the third day the dragoman fell ill, and Hemprich had to rely fully on Franz Kreysel, a German servant and guard who had joined him in Cairo. In Sanhur near Lake Qarun a suitable spot was found from which to conduct research,

Hand-coloured lithograph of scorpions.
Christian Gottfried Ehrenberg, *Symbolae physicae seu Icones et descriptiones corporum naturalium novorum aut minus cognitorum quae ex itineribus Libyam Aegyptum Nubiam Dongalam Syriam Arabiam et Habessiniam …*, Pars Zoologica II, Arachnoidea, Berlin 1828, plate II
Linnean Society Library, FF 916.1/.3:57 EHR

22. A.V. Abramov and G.F. Baryshnikov, 'Geographic Variation and Intraspecific Taxonomy of Weasel *Mustela nivalis* (Carnivora, Mustelidae)', *Zoosystematica Rossica*, 8, 1999, pp. 365–402.

23. Hemprich to Lichtenstein, 23 April 1821, Streseman, *Hemprich und Ehrenberg*, p. 33.

24. 'Meine Lage wahrlich keine angenehme. Allein nun mit aller Arbeit beladen, von Pest umgeben, ohne Nachricht von Berlin […] Sollt' ich nicht wiederkehren, nicht wahr, Sie sprechen ein Wort für meine Mutter?' Hemprich to Lichtenstein, 5 May 1821, Stresemann, *Hemprich und Ehrenberg*, p. 35.

25. 'Prof. Liman ist dahin, der General ist ziemlich unverrichteter Sache auf dem Rückweg, und wenn wir auch umkehren, so ist die ganze große Expedition = Null.' Hemprich to Lichtenstein, 30 May 1821, Stresemann, *Hemprich und Ehrenberg*, p. 38.

but when Hemprich returned from a first exploration of the area, he found that the dragoman had been shot at during his absence and that the governor of the town ordered them to stay in his house because it was not safe to bivouac in the wild. Reluctantly Hemprich accepted. Ehrenberg arrived unexpectedly, exhausted and suffering from ophthalmia. The Arab to whose care he had been entrusted had cheated him out of his money, so that he had no choice but to leave Cairo and join Hemprich. When Kreysel fell ill and died, they decided they had all better return to the capital.

Although the Fayyum project proved highly frustrating to Hemprich and Ehrenberg, they still managed to send to Berlin over a hundred bird skins of species shot around Saqqarah, as well as 3,600 insects, 1,000 plants, 86 mammals, and 153 birds collected in the Fayyum. In both areas they observed the Egyptian Jackal, which Ehrenberg placed as a subspecies in Linnaeus' species of the Golden Jackal, *Canis aureus lupaster*, in 1832.

As Ehrenberg and Hemprich were waiting for funds from Berlin, an opportunity to continue their research suddenly came up. The viceroy and pasha of Egypt, Muhammad Ali, was bent on expanding Egyptian dominion in the Sudan, and needed gold to finance his miliary campaigns.[26] In 1821 he wanted an expedition to go south into Dongola to examine the site of a recently discovered gold mine, and, on the Prussian consul Buccianti's advice—he had recommended the naturalists as able mineralogists—the two men were invited to join. Although the Prussian government had asked them not to waste time (and money) in Upper Egypt and give the Red Sea area priority, they felt they could not refuse. The area from Nubia into Dongola and Darfur was relatively unexplored territory—it had been reported that there were lions, panthers, gazelles, hippopotamuses, snakes, scorpions and huge spiders to be caught there—and to Hemprich there was no doubt that the Red Sea had to wait. To Lichtenstein he expressed his hope that, after much misfortune, the time of success had finally come.[27] Ehrenberg, who had recovered considerably and started to look forward to being active again, fully shared his friend's eagerness and in a letter to Nees von Esenbeck could not hide his enthusiasm: 'What should we do? Should we stay in Cairo for 6 months blocked by the flood? Definitely not! We'll south and to Dongola! No botanist has seen a Darfur spring yet. Many a gentleman wishes us luck, the Heavens will not be against us forever!'[28] When the pasha finally changed his mind and abandoned the idea of the expedition into Dongola, the naturalists decided to undertake the journey by themselves.

Dongola

In August 1821 Ehrenberg followed in Geoffroy, Savigny and Delile's trail. Exactly 22 years after Napoleon ordered the French naturalists to explore upper Egypt, he

26. See Afaf Lutfi al-sayyid Marsot, *Egypt in the Reign of Muhammad Ali*, Cambridge 1984, p. 205.
27. Hemprich to Lichtenstein, 2 August 1821, Stresemann, *Hemprich und Ehrenberg*, p. 39.
28. 'Was thun? Sollen wir 6 Monate in Cahira von der Fluth umgeben sitzen? Gewiß nicht! Auf denn gen Süd und Dongala! Ein Frühjahr in Dar-foor hat kein Botaniker gesehn. Glück auf wünscht uns so mancher Ehrenmann, der Himmel wird nicht immer auf uns zürnen!' Ehrenberg to Nees von Esenbeck, 10 August 1821, Stresemann, *Hemprich und Ehrenberg*, p. 44.

rented a boat and set sail for Beni Suef, where Hemprich, who had stayed behind in Alexandria to supervise the shipment of the Fayyum collection to Berlin, would join him. Even if Ehrenberg still felt rather weak during most of September, he did a great deal of collecting on the way, and in Beni Suef he prepared over a hundred bird skins and almost a thousand insects. Like Forsskål and Geoffroy before him, he had natives, and children in particular, bring him all kinds of insect and reptile, giving them small rewards or *bakshish* when they brought in something useful. Although this initially worked well, it soon got the naturalist into trouble with the local population. Speculating on why he should be so much interested in insects and snakes they surmised that what he was really after was their children, whom he caught in order to extract poison from their bodies. As some children were reported missing by their mother, the local bey sent soldiers to investigate the contents of the crates Ehrenberg was preparing. The naturalist did not allow the soldiers to touch his collections, very likely remembering the fate of Forsskål's collections in Mocha, but invited the bey to come and see for himself. Luckily the children, who had been playing in a shop, were found and the bey decreed that the next person who falsely accused 'den fremden Ärzten' would be thrown into the Nile. Although the hostility of the local population was thus averted, Ehrenberg admitted to Link that the affair could have ended very differently, as the situation was similar to the one which had cost Ulrich Jasper Seetzen his life in the Yemen.[29]

When Hemprich had finished his work in Alexandria and had joined Ehrenberg again, they proceeded to Aswan, which they reached in the first half of November. Here, before taking a smaller boat past the cataracts, they rented storage room to deposit 450 birds, 4,500 insects and 20 packets with pressed plants. These they would pick up on their return journey to Cairo. In his last letter from the Egyptian border Hemprich drew attention to what he thought were an unknown weasel species ('*Memphitis*') and a large bat species ('*Vespertilio*'),[30] but there was no time to examine the characteristics of these mammals in more detail. Amongst the birds he caught he also mentions a new nightjar. When studied at ease in Berlin years later they indeed turned out to be new species. The weasel species was the Saharan Striped Polecat (*Ictonyx libyca*), and the Desert Long-eared Bat was named by Wilhelm Peters after its discoverer as *Otonycteris hemprichii* in 1859. The nightjar was described by Lichtenstein in 1823 as *Caprimulgus aegyptius*.

Travelling past the cataracts was not without problems. Of the five European servants they had engaged, three fell ill before leaving Aswan, and they had to go to and fro carrying their goods past the cataracts. Beyond the cataracts at Aswan Ehrenberg found transport by boat to Wadi Halfa while Hemprich followed with the camels. From there to Dongola they had to travel overland with their exhausted camels. But it was all worth it. When they reached the town in February 1822 they were greeted by the local bey with unexpected hospitality. Abidin Bey, who had been treated well by the French when they invaded the country, had a friendly disposition to all Europeans, and he generously provided Ehrenberg and Hemprich with lodgings. He insisted on paying all their expenses and gave them a military

29. Ehrenberg to Link, 27 October 1821, Stresemann, *Hemprich und Ehrenberg*, p. 47. Seetzen was a German traveller who was poisoned in the Yemen because the local population thought he was a magician.
30. Hemprich to Lichtenstein, 10 November 1821, Stresemann, *Hemprich und Ehrenberg*, p. 48.

August-Friedrich Schmidt, hand-coloured lithograph of the Short-tailed Crimson Weaver from Dongola.
Christian Gottfried Ehrenberg, *Symbolae physicae seu Icones et descriptiones corporum naturalium novorum aut minus cognitorum quae ex itineribus Libyam Aegyptum Nubiam Dongalam Syriam Arabiam et Habessiniam* ..., Pars Zoologica I, Avium, Berlin 1828, plate II
Linnean Society Library, FF916.1/.3:57 EHR

escort on their excursions around the town. For the first time the naturalists could work without being hampered by lack of money or being disturbed by a hostile local population. And for the first time there was a feeling of satisfaction: 'The liveliness of the heart of Africa forms pretty groups of people around us and we begin to feel pleasure in our travels,' Ehrenberg wrote to Rudolphi.[31]

In the Province of Dongola the two naturalists were able to collect an enormous number of plants, animals and minerals, which were to form by far the largest collection they would send back to Germany. The numbers of specimens are truly astonishing. On 1 October 1822 Hemprich forwarded to Berlin 10,000 insects, 1,622 birds, 233 mammals, 242 reptiles and three chests with pressed plants, all collected between mid-September 1821 and mid-July 1822, including the specimens they had stocked at Aswan. Lichtenstein too seemed bewildered by the sheer amount of material arriving at the museum in Berlin, and warned the two naturalists that, when they returned to Germany, they would have to work hard.[32] In the letters of this period both naturalists wrote with enthusiasm about their numerous finds. They were frequently unable to identify the animals with precision and suspected new species. There were unknown dogs, bats, jirds, falcons, pigeons, sandgrouse, ducks, bee-eaters, hummingbirds, water-beetles and catfish. But as there was no time to classify them with certainty, these are often simply mentioned (and labelled) as new species, although Latin names were occasionally coined for them.

From the beginning of their expedition to the East Hemprich had been much interested in gazelles, and, if he had not been very successful in shooting the animals in Lower Egypt, he had no difficulty now in getting a good collection of specimens. He even believed they had found a new species. And on 10 May he proudly announced the first new genus of bird, which, in honour of the hospitable bey of Dongola, he proposed to call *Sphenorhynchus Abdimii*. The collection of insects grew out of all proportion, and they had to remove duplicates to make space for the numerous new species. On the other hand, they were disappointed with the very few plants they found, and although some were unknown to them, at the beginning of August Ehrenberg could only conclude that, 'for a botanist, the Nile was an extremely boring river'.[33] But when the rain came two weeks later he witnessed for the first time the richness of the north African flora. The description of this event in a letter to Nees von Esenbeck brought out the poetic effect this had on the otherwise prosaic naturalist.[34] Ehrenberg could now boast that his collection also included a great number of new plant genera.[35]

31. 'Das Lebendige des Herzens Afrikas bildet hübsche Gruppen um uns und wir fangen an Freude an unserer Reise zu haben' Ehrenberg to Rudolphi, 2 April 1822, Stresemann, *Hemprich und Ehrenberg*, p. 56.

32. Lichtenstein to Hemprich and Ehrenberg, 5 October 1822, Stresemann, *Hemprich und Ehrenberg*, p. 60.

33. 'Der Nil ist für einen Botaniker ein äußerst langweiliger Fluß.' Ehrenberg to Link, 3 August 1822, Stresemann, *Hemprich und Ehrenberg*, p. 62.

34. Am Abend noch sah ich mit traurigem Sinn/ Auf dürre sandige Steppen hin,/ Da strömte der Regen vom Himmel flugs,/ Da schwollen die Bächlein, der Nilstrohm wuchs,/ Da träumt mir's von hellen Flamm' in der Nacht,/ Und als ich am frühen Morgen erwacht,/ Da sah ich die Flora in herrlicher Pracht/ Da hat mir das Herz in dem Busen gelacht. [In the evening I still looked with a sense of sadness/ Over the parched sandy steppe,/ There the rains poured down from the skies,/ There the stream swelled, the Nile-flood rose,/ There I dreamt of the bright flame in the night,/ And as I woke early in the morning,/ I saw the flora in exquisite splendour/ And it made all my heart laugh out.]

35. Ehrenberg to Nees von Esenbeck, 20 August 1822, Stresemann, *Hemprich und Ehrenberg*, p. 67.

Hinrich Lichtenstein

August-Friedrich Schmidt, hand-coloured lithograph of *Nectarinia metallica.* Christian Gottfried Ehrenberg, *Symbolae physicae seu Icones et descriptiones corporum naturalium novorum aut minus cognitorum quae ex itineribus Libyam Aegyptum Nubiam Dongalam Syriam Arabiam et Habessiniam* ..., Pars Zoologica I, Avium, Berlin 1828, plate I Linnean Society Library, FF 916.1/.3:57 EHR

When and by whom were all these new species ultimately described and published? The standard publication of Hemprich's and Ehrenberg's work in the Near East is *Symbolae physicae, seu Icones et Descriptiones corporum naturalium novorum aut minus cognitorum, quae ex itineribus per Libyam, Aegyptum, Nubiam, Dongolam, Syriam, Arabiam et Habessiniam* ... Started in 1828, it was originally envisioned as the German counterpart of the natural history volumes of the *Description de l'Égypte*, coming out in unbound folio instalments of Latin text with fine coloured plates in a separate portfolio. Most of the plates, prepared from Ehrenberg's drawings, were printed and coloured by hand under his supervision in 1828–9. They were subsequently published with the text parts between 1828 and 1834, after which he lost interest in the project. Admittedly, with the help of others, new parts kept coming out for the rest of the century, some of them posthumously—the original plates of the 1820s for the Pars Botanica, for example, were issued for the first time in 1900 with a text in German by Carl Moritz Schumann—but the great project of describing the natural results of the expedition remained unfinished.

The reasons why Ehrenberg left *Symbolae physicae* unfinished are threefold. First there was Hemprich's premature death. Next, Ehrenberg's scientific interest shifted sharply to the field of micro-organisms, and his greatest contribution to science was the classification of Infusoria, first presented in 1838 in his *Die Infusoriensthierchen als vollkommene Organismen*. Finally, Ehrenberg's refusal to continue *Symbolae physicae* had much to do with what Lichtenstein had done to the extensive collection of animals and plants he and Hemprich had sent at regular intervals to Berlin.

Lichtenstein was the son of the naturalist and professor of oriental languages Anton August Lichtenstein, who produced a series of sales catalogues of important private collections of mammals, birds, shells and insects that were to go on auction during the 1790s. In these catalogues the objects were listed according to the new Linnaean system of nomenclature with descriptions in German, except for new taxa which were described in Latin. This gave many of the new items taxonomic validity.

When, at the end of December 1821 Hinrich Lichtenstein wrote to Ehrenberg and Hemprich about his attempts to raise the funds needed to meet their constant demands for money, he urged them to collect as many samples of animals as possible 'as the expenses of your travels will subsequently be paid for, as far as possible, by the sale of these'.[36] The naturalists are unlikely not to have agreed with this plan, but if they did feel uncomfortable about it the specification *nachmals* ('afterwards') would have reassured them. Nothing could be advanced against the idea of collecting duplicates which could be sold to recover the costs of the expedition once they had sorted their material after their return to Germany, especially as the Instruction of the Academy had decreed that, although they were asked to communicate their observations, 'publication of these notices would remain their sole right'.[37] There

36. '[...] weil die Kosten Ihrer Reise nachmals aus dem Verkauf derselben soviel wie möglich erstattet werden sollen.' Lichtenstein to Hemprich and Ehrenberg, 20 December 1821, Stresemann, *Hemprich und Ehrenberg*, p. 50.

37. '[...] die öffentliche Bekanntmachung dieser Anzeichnungen ihnen allein vorbehalten bleiben soll.' Stresemann, *Hemprich und Ehrenberg*, p. 14.

was hardly any reason for Ehrenberg and Hemprich to suspect Lichtenstein of unfair dealings at this stage. However, in retrospect one might wonder why Lichtenstein felt it necessary to remind the naturalists in a following letter that their collections would be of great scientific value and 'that science will significantly benefit from them, once you have described and published everything after your return, for which event I am preparing with all my strength, without disclosing in the least your discoveries in untimely publications'.[38] One also wonders how the two naturalists received the news that by 20 June 1823 Lichtenstein had 'completely worked through' their collection from Upper Egypt, Nubia and Dongola, and whether they suspected anything when he enumerated many of their new species, as, for example, in the case of Hemprich's new bird genus: '*Sphenorhynchus Abdimii* is really a *Ciconia* [...] I identified as a new species and therefore gave it a name, which I have exchanged for yours: *Ciconia Abdimii*.'[39]

Did Hemprich and Ehrenberg ask themselves what Lichtenstein meant by 'I identified' and '[I] gave it a name'? Again, Lichtenstein's repeated complaints about lack of space at the Museum for the new collections sound alarming in retrospect. The collection 'is growing to amazing—and *frightening*—dimensions,' he concludes a letter in 1822.[40] Did Hemprich and Ehrenberg ever realize what Lichtenstein's collective statements amounted to? These were the facts: there was no money for their return journey, there were requests for duplicate samples which were meant to be sold, there was a chronic lack of space at the museum, and Lichtenstein reported on their collections with a startling emphasis on *his* classification of *their* species. Could they not foresee what was going on in Berlin? It is hard to believe they did not. Yet there is no evidence of this in the letters. It was only after Ehrenberg's return to Germany that he discovered that the integrity of their collection had been seriously tampered with. Apart from the fact that original labels had been removed or misplaced when the boxes were opened at the Museum, part of the collection had disappeared through a reckless policy of selling and exchanging items. Even unique specimens had been heedlessly sold as duplicates. When Ehrenberg discovered that 'much has been published under the names of others', he had every right to complain that by this selling and exchanging the sole rights to publish the results of their travels had been pre-empted.[41] Lichtenstein's own professional ethics in this matter were not entirely spotless either. In 1823 he had published a sales catalogue of the Museum's so-called duplicates, describing the items for sale in Latin, as his father had done before him. In this manner a number of Hemprich's and Ehrenberg's new species were validly named and described by Lichtenstein

38. '[...] von großem Wert für die Wissenschaft und wird derselben bedeutenden Nutzen bringen, wenn Sie nach Ihrer Rückkehr alles beschreiben und der Welt bekannt machen, wozu ich nach Kräften alles vorbereite, ohne Ihnen im mindesten durch unzeitige Publication vorzugreifen.' Lichtenstein to Hemprich and Ehrenberg, 23 April 1822, Stresemann, *Hemprich und Ehrenberg*, p. 53.

39. '*Sphenorhynchus Abdimii* ist eine wahre *Ciconia* [...] ich erkannte sie für neu und gab ihr einen Name, den ich aber nun mit dem Ihrigen vertauchscht habe: *Ciconia Abdimii*.' Lichtenstein to Hemprich and Ehrenberg, 20 June 1823, Stresemann, *Hemprich und Ehrenberg*, p. 80.

40. My italics. 'Es wächst jetz zum Erstaunen und—Schrecken.' 5 October 1822, Stresemann, *Hemprich und Ehrenberg*, p. 60.

41. 'Vieles ist schon unter fremdem Nahmen publicirt'; 'Durch diesen Verkauf und Tausch ist jenes uns zugestandne Recht der alleinigen eignen Publication der Reise-Resultate inhaltslos geworden.' Ehrenberg to Altenstein, 3 January 1827, Stresemann, *Hemprich und Ehrenberg*, p. 166.

long before the two naturalists could return to Germany to take the proper credit for these discoveries.[42]

Cairo and Damietta

The expedition to Nubia and Dongola had been so successful that the two naturalists soon ran out of chests to store and transport their collections. In August 1822 Hemprich decided to travel to Alexandria with their treasure of 22 boxes of skeletons, bird skins and insects, while Ehrenberg wanted to stay a bit longer and continue collecting with the help of a few servants. However, Ehrenberg's exultant mood in Dongola was short-lived. On 17 September illness struck again and malaria and eye infection incapacitated him for two months. 'Our journey seems to be star-crossed,' he wrote to Nees von Esenbeck.[43]

Hemprich also met with disappointment. In Cairo he found a letter from Lichtenstein awaiting him, whose content and tone came as an unpleasant surprise. Expecting acknowledgement and thanks for the early results of the expedition and full future financial support, Hemprich read with increasing anger that the Prussian state demanded instead a detailed account to justify all their expenses, and ordered them back to Germany as soon as they returned from Upper Egypt. In his reply Hemprich did not hide his chagrin:

> *After thousands of efforts, risk of life and our utmost endeavours to satisfy the Academy and the Ministry, these are our thanks. I admit that I had not expected a letter from your hand, and with your signature, which so deeply offends and unfairly accuses us.*
>
> [...] *In short, I am afraid that there has been no confidence in our work. We will not lose by it, it is the collections and science that lose by it, and the Italians will laugh:* quie [*sic*] brutti Tedeschi non son capaci di finir' un viaggio [*those horrible Germans cannot even bring a voyage to an end*].[44]

He next wrote to Ehrenberg that their expedition had come to an end, but he did not mention the ministry's requests, as 'it would have killed him'. But this letter would only reach his friend in Nubia on 22 December. Although by the time Ehrenberg joined Hemprich in Cairo in January 1823 word of hope had come from

42. The descriptions of new species collected in Upper Egypt and Nubia that thus received taxonomical validity in *Verzeichnis der Doubletten des zoologischen Museums der Königliche Universität zu Berlin* include: *Ammomanes deserti*, *Anthreptes metallicus*, *Caprimulgus aegyptius* *Caprimulgus nubicus*, *Ciconia abdimii*, *Cypsiurus parvus*, *Emberiza striolata*, *Lanius nubicus*, *Oenanthe lugens*, *Passer luteus*, *Passer simplex*, *Prinia gracilis*, *Pterocles coronatus*. The sales catalogue also described various new species collected in Lower Egypt.

43. 'Es scheint ein trüber Stern über unsrer Reise zu walten.' Ehrenberg to Nees von Esenbeck, 22 October 1822, Stresemann, *Hemprich und Ehrenberg*, p. 67.

44. 'Nach tausendfacher Mühe, Wagniß des Lebens und möglichstem Streben der Akademie und dem Ministerium genug zu thun ist dies der Dank. Ich gestehe ich hätte dies nicht erwartet, am allerwenigsten hätte ich ein Schreiben, das uns im Innersten verletz und höchst ungerecht tadelt, von Ihnen unterzeichnet, von Ihnen verfaßt erwartet. [... K]urzum es ist zu bedauern, daß es an Zutrauen zu uns gefehlt hat. Wir verlieren nichts, es sind die Sammlungen und die Wissenschaft welche verlieren, und die Italiäner sagen lachend: *quie [sic] brutti Tedeschi non son capaci di finir' un viaggio*.' Hemprich to Lichtenstein, 16 September 1822, Stresemann, *Hemprich und Ehrenberg*, pp. 69–70.

Lichtenstein, when Ehrenberg heard they had been asked by the government to account for their expenses he found it necessary to speak his mind to Lichtenstein: 'Instead of friendly sympathy and recognition from our country, we found unfair, truly unfair, reproaches and slanderous letters.'[45] And the very same day he wrote with cutting irony to the President of the Academy, Nees von Esenbeck: 'We should have stayed in Cairo, shot 1,000 specimens of *Charadrius aegyptius* and thus recovered all the money.'[46]

Most of the spring and early summer of 1823 the naturalists waited for decisions from Berlin. In April they pitched their tent in the surroundings of Damietta for a couple of days to study the flora and fauna of Lake Manzalah. Ehrenberg still had bouts of malaria. Quinine, of which he had already consumed about 'half a shipload', helped, but every time he got wet the fever returned. However, this did not prevent him from wading in the shallow water in search of unknown species. Hemprich caught a new reed warbler (*Acrocephalus stentoreus* Hemprich & Ehrenberg), and Ehrenberg collected a good number of plants. When a rebellion against the Kashef broke out in Damietta, work became dangerous. Hemprich was dissecting fish one day onboard with two pistols and a sabre at hand, and Ehrenberg, who had gone botanizing, had to make his way back to the vessel with an uncocked rifle in one hand and a rare ranunculus in the other.

At the beginning of May they were back in Cairo. As still no news about funding was forthcoming, Hemprich and Ehrenberg informed Lichtenstein and Nees von Esenbeck that they had decided to use the remaining sum of money for an excursion to the Red Sea and come back to Cairo towards the end of September to see whether new orders from Berlin had come in. However, just before leaving, news (and money) came, and the trip to the Red Sea was now turned into a major expedition from which they would not return until the spring of 1825. Besides Martin Bretzka, the German who had travelled with them in Nubia and Dongola, they employed two other Germans as hunters.

The Red Sea

Like Forsskål half a century before him, Ehrenberg did not find much of botanical interest at Suez, a town which he regarded as one of the most miserable places on earth.[47] It had not rained for months, and even insects were difficult to find. Two of the hunters were posted for gazelles, while the third went shooting birds. Hemprich, who managed to rent a boat with a crew with difficulty, 'feasted' on the wealth of sea animals. Most of the fish he caught had been described by Forsskål, but he suspected many novelties among the thousands of sea-slugs, crustaceans and sea-centipedes he collected.

45. 'Statt freundlicher Theilnahme und Anerkennung aus dem Vaterlande fanden wir unverdiente, wahrlich unverdiente Vorwürfe und entehrende Briefe.' Ehrenberg to Lichtenstein, 6 February 1823, Stresemann, *Hemprich und Ehrenberg*, p. 78.

46. 'Wir hätten sollen in Cahira bleiben, 1000 Stück von *Charadrius aegyptius* schießen and so das Capital wieder einbringen.' Ehrenberg to Nees von Esenbeck, 6 February 1823, Stresemann, *Hemprich und Ehrenberg*, p. 81.

47. Ehrenberg to his father, 29 June 1823, Stresemann, *Hemprich und Ehrenberg*, p. 91.

An interesting discovery in Suez was an unknown shrew, which, in a letter to Rudolphi, he called *Sorex crassicaudus*.[48] The name was eventually published by Lichtenstein in 1829.[49] In 1832, however, Ehrenberg decided to place the animal as a subspecies *sacer* of Linnaeus' *muriates*, but at the same time made it the genotype of a new genus: *Suncus*. Unlike Lichtenstein, Ehrenberg was right to recognize that the animal described was not a new species but a specimen of Linnaeus' Asian House Shrew or Asian Musk Shrew, an Indian species which can be highly variable in size and colour. The fact that Hemprich caught the animal in Suez, far from its original Indian habitat, might have been misleading at first. But the Asian House Shrew easily adapts to environmental change and is successful in spreading to new geographical areas, and may have been introduced in Egypt through human agency, either unintentionally with cargo from India, or purposely to keep away rats and snakes from habitations. Today, Hemprich's original name, published by Lichtenstein, is deemed invalid, as is Ehrenberg's and Hemprich's subspecies *sacer*. However, the new generic name proposed in *Symbolae physicae* is still the accepted name.

In Tur Ehrenberg was shocked by the 'naked and lazy people' who, apart from a bit of fishing, stayed at home with their wives to sleep away their time. Compared to Damietta, collecting plants and animals was quite easy now. During their excursions inland they were no longer pestered by the local population since, by the end of 1823, they spoke enough Arabic to look 'truly sullen' (*bärbeissig*) with their guns and pistols. Sailing down the Red Sea in their small boat was dangerous, however, and they were in constant fear of getting drowned. Moreover, they felt uncomfortably cramped in the small hold and suffered from sea-sickness. But there was no money to rent a bigger vessel. The food too was bad. There was no meat to be had and they subsisted on fish, rice, lentils and bread of very poor quality. What was worse, fresh water was not to be had either. 'One needs a lot of resignation and courage to be able to accept our situation,' Hemprich admitted to Rudolphi.[50]

On the other hand, they could not deny the success of the Red Sea expedition as naturalists. Ehrenberg wrote to his father that they came across many unexpected items on these coasts, almost as much as during three years in Egypt.[51] Ehrenberg and Hemprich were constantly aware that their finds would be measured against Forsskål's and Savigny's imposing collections in Europe, and by the end of August they boasted that they had found 160 species of fish, including about 60 'Forsskålean' species, as well as nearly all Savigny's annelids.[52] The greatest revelation of all was perhaps the wealth of material they collected on their excursions from Tur to Mount Sinai, which the naturalists of both the Danish and French expeditions had failed to explore. In September 1823 Hemprich started to explore Mount Sinai alone, while Ehrenberg, who had injured his foot and was unable to travel, stayed in Tur to sort, describe and pack their collections. It had been planned that Hemprich would accompany their new collection to Alexandria as soon as he returned and supervise

48. Hemprich to Rudolphi, August 1823, Streseman, *Hemprich und Ehrenberg*, p. 92.

49. 'Über die afrikanischen uns asiatischen Arten der Gattung *Sorex*', *Verhandlungen der Gessellschaft Naturforschender Freunde Berlin*, p. 386.

50. 'es gehört viel Resignation und Muth dazu unsre Lage erträglich zu finden.' Hemprich to Rudolphi, undated, Stresemann, *Hemprich und Ehrenberg*, p. 93.

51. Ehrenberg to his father, 9 July 1823, Stresemann, *Hemprich und Ehrenberg*, p. 91.

52. Hemprich to Rudolphi, undated, Stresemann, *Hemprich und Ehrenberg*, p. 92.

the despatch of the chests to Berlin. When he came back from Mount Sinai with glowing reports and a wealth of new material, however, Ehrenberg decided to go up himself by a different way, adding considerably to Hemprich's finds. Ehrenberg reported to Nees von Esenbeck that nearly all the plants he came across on the eastern road to Mount Sinai were new to him, and that of the 120 plants collected in the peninsula so far he had only recognized 48. He suspected the existence of at least nine new genera and 63 new species.[53] 'What the Red Sea is for zoology, those mountains are for botany,' Ehrenberg declared in a letter to Link.[54]

On 16 November Hemprich dispatched a collection from Alexandria which almost rivalled the collection from Upper Egypt. It contained about a hundred mammals, over a thousand birds, 1,600 shells, 700 fishes, 330 worms, 300 corals, as well as numerous specimens of amphibians, reptiles, and insects. There were also 42 packets of pressed plants. Apart from Red Sea material these numbers also include specimens collected by Ehrenberg during his stay in Ambukohl in the summer of 1822, and the specimens collected around Damietta during the spring of 1823.

It was often difficult to obtain undamaged animals for the museum in Berlin, especially when the local population was employed to bring in specimens. Birds and mammals were frequently badly shot and fish were smashed to death. In this condition specimens were only of limited value for research, while they were totally worthless as duplicates that the museum could sell. It was only when they stayed in a place for some time, as, for example, in Tur, that they could train (*abrichten*) the people to find what they were looking for. In November, while Hemprich was arranging the shipping of material from Alexandria, Ehrenberg managed to obtain perfect skeletons of what he called *Hyrax syriacus*. Rock Hyraxes (the coney of Leviticus 11:5 and Deutoronomy 14:7),[55] though resembling rabbits, are small hoofed animals related to elephants. The animal, common in the drier rocky areas of Africa and Arabia, was known in Europe from a species described from the Cape (*Hyrax capensis*) by Pallas in 1766. The German naturalist Johann Christian Daniel von Schreber had described a hyrax from the Near East in 1784, but it remained debatable whether this species was indeed distinct from the Cape species. Hemprich, who had caught the animal during their excursions in Upper Egypt and Dongola, now became convinced that there were three different hyrax species: the Dongola species, he argued, was different from the Cape one, and the Sinai species near Tur was clearly different from the Dongola one. The first was given the name *H. ruficeps* in 1828, while for the species on Mount Sinai Schreber's *H. syriacus* was adopted. Yet another species from Eritrea was described in 1828 as *H. habessinicus*. The new species from Dongola and *H. syriacus* are depicted in the *Symbolae physicae*, but, strangely enough, the new Eritrea species is not. Although in the text description of *H. syriacus* Ehrenberg tentatively adds the name *sinaicus* between brackets, following Hemprich's manuscript indication, he does not do this in the heading, which only specifies 'Hyrax *syriacus* / E MONTE SINAI', and he nowhere attempts to contrast it with Schreber's species. Instead, he is only concerned with establishing *H. syriacus*

53. Ehrenberg to Nees von Esenbeck, 4 October 1823, Stresemann, *Hemprich und Ehrenberg*, p. 94.

54. 'Was das rothe Meer für Zoologie war, war jenes Gebirge für die Botanik.' Ehrenberg to Link, 12 October 1823, Stresemann, *Hemprich und Ehrenberg*, p. 96.

55. See also Proverbs 30:26, Psalm 104:18.

as distinct from *H. capensis*. Today the taxonomical status of *Hyrax* species is still a controversial question. Lumpers consider all of them as synonyms of the original Cape Hyrax, whose modern name is *Procavia capensis*. Others, however, accept them as subspecies. There seems to be even less consensus about the status of Hemprich and Ehrenberg's 'sinaicus'. Although John Edward Gray picked up the name *H. sinaiticus* in 1868, David Harrison in *Mammals of Arabia* (1968) included it in *Procavia capensis syriaca*, while Menache and Ferguson in *The Mammals of Israel* (2002) stress that it is a morphologically, geographically and ecologically distinct form which needs to be considered as a subspecies in its own right: *P. capensis sinaiticus*.

On 15 December Ehrenberg wrote to Hemprich, who was still in Rosetta, that he had obtained two gazelles which he believed belonged to a new species, but he did not seem much impressed with the discovery. The most important find of the last few days, he thought, were thirteen new genera of nereids, and he added in his letter that 'two new species of infusoria and the thing which every now and then makes the sea quite red interest me most.'[56] The 'thing' which occasionally cast a red-brownish glow over the Red Sea was a planktonic alga whose dying blooms cause a reddish colour on the surface of the water. The specimens of this alga which Ehrenberg collected at Tur were the basis of his description of this cyanobacterium in later years,[57] and of its current name, *Tricodesmium erythraeum*. Thus, half a century after the Danish expedition, Ehrenberg found a definitive answer to Michaelis' question about *suph*, with which his *Fragen an eine Gesellschaft gelehrter Männer* opened.[58]

Hemprich was clearly more interested in the gazelles Ehrenberg had forwarded to him in Alexandria, and which he marked *Antilope arabica*, a name which has provoked some confusion. In his letter of 15 December Ehrenberg mentioned two gazelles, while on 3 January he wrote with surprising imprecision that he had sent Hemprich '2–3 Antilopen'. On 16 January he reported again that he had obtained the skeleton of a male gazelle.[59] Although the two specimens mentioned in the letter of 15 December are well-known to zoologists, the references in the later two letters seem to have escaped attention. This is partly because only the skull and skin of an adult female and the skin of a juvenile female[60] are today in the collection of the Berlin Zoological Museum,[61] while the skin and skull of a third specimen in the Berlin collection came from an animal from the Farasan Islands, just off the coast of southern Arabia, which the naturalists visited more than a year later.[62] These specimens were lumped together by Ehrenberg in the description of *A. arabica* in

NEXT OPENING
Friedrich Bürde, hand-coloured lithograph of 'Antilope Arabica. Sinai' (*Gazella dorcas isabella*). Christian Gottfried Ehrenberg, *Symbolae physicae seu Icones et descriptiones corporum naturalium novorum aut minus cognitorum quae ex itineribus Libyam Aegyptum Nubiam Dongalam Syriam Arabiam et Habessiniam* ..., Pars Zoologica I, Berlin 1828, plate V
The Arcadian Library, 10503

56. Ehrenberg to Hemprich, 15 December 1823, Stresemann, *Hemprich und Ehrenberg*, p. 105.
57. Christian Gottfried Ehrenberg, 'Neue Beobachtungen uber blutartige Erscheinungen in Aegypten, Arabien und Sibirien', *Annalen der Physik und Chemie*, 8, 1830, p. 506.
58. See above, chapter 1.
59. '[E]ine mänliche Gazelle als Skelet'. The reference to the skeleton in this letter is of potential interest. However, there is no evidence that it ever reached Berlin, nor that it was lost on the way. Ehrenberg and Hemprich often had to discard collected material because of a lack of space in the crates, but it is unlikely they would have left out their only complete skeleton.
60. Cf. Colin Groves 'Notes on Gazelles: IV. The Arabian Gazelles Collected by Hemprich and Ehrenberg', *Zeitschrift für Saugetierkunde*, 48, 1983, p. 321.
61. Oscar Rudolph Neumann discarded the Sinai specimens as types for *Gazella arabica* and described them as *Gazella arabica ruepellii* instead. Groves, however, regards this subspecies as a synonym of *Gazella arabica isabella* Gray (1846).
62. Ehrenberg to Lichtenstein, 30 April 1825, Stresemann, *Hemprich und Ehrenberg*, p. 141.

Symbolae physicae (1828): 'ex vallibus sinaiticis, arabia deserta littorali et ex insula insula maris rubri Farsan, Arabiae Felices urbi Gisan proxima'. Lichtenstein, too, who received the Sinai specimens in April 1825 and the Farasan specimen in April 1826, used them all indiscriminately to describe the species in 1827.[63] Ehrenberg probably further contributed to the confusion when he stressed the existence of the species *dorcas* and *arabica* in both the Sinai and in Syria. The uncertain status of Hemprich's gazelle during the nineteenth century is reflected by Edward William Lane, a linguist rather than a scientist, who, in his *Arabic–English Lexicon* (1863), says that it is 'a variety [of *Gazella dorcas*]; or, accord. to some, each is a distinct species of gazelle' (I.1908). Today *Gazella arabica* is an accepted name, but of an enigmatic entity. Colin Groves has shown that the lectotype of *G. arabica*—Hemprich's and Ehrenberg's skull and skin of the male specimen from the Farasan islands—differs from all other known gazelles and must therefore be seen as a separate species, for which the name *arabica* needs to be adopted.[64] The distinctiveness of the taxon has convinced most specialists. The odd thing about this species, however, is that no other specimens of this entity have ever been found. It is on such grounds (distinctiveness and lack of further information) that the *IUCN Red List of Threatened Species* considers *G. arabica* an extinct species.[65] In 1985, however, Groves and Lay described a new species from the Yemeni highlands as *G. bilkis*, and noted that it resembled to a great extent the characteristics of the specimen of *G. arabica* from the Farasan islands. In 1997 Groves took the step of uniting the two species, making the Yemeni animals a subspecies of Ehrenberg and Hemprich's type: *G. arabica bilkis*.[66] Seeing the long controversy over *G. arabica,* it is a pity that the animal depicted in *Symbolae physicae* is a specimen of *G. dorcas isabella*, based on Ehrenberg's Sinai specimens of December 1823.

Before returning home in October 1825 there were still plans to travel further south to Jiddah, Sawakim and Mocha, and then to try Kordofan (Sudan) once more in the spring. But for this the naturalists needed approval (and money) from Berlin. It was highly frustrating that no letters should have come—'It is unforgivable that people in Trieste or Berlin ruin such an expensive journey [...] We are wasting the best part of our youth desiring what we cannot have'[67]—and fatigue was making itself felt. Hemprich told Ehrenberg that by the end of 1824 he wanted to be back in Europe, that he was willing to go once more into the heart of Sudan, but that he would not join his friend on a journey to Mocha and Abyssinia.[68] Ehrenberg, on the other hand, had lost all interest in Kordofan. While Hemprich was waiting for instructions in Alexandria, he indefatigably continued collecting along the Red Sea, and now more than ever before was convinced that their discoveries here would be the crowning achievement of their expedition. On 3 January 1824 he

63. Lichtenstein, *Darstellung neuer oder wenig bekannter Säugethiere*, Berlin 1827.

64. Groves, 'Notes on Gazelles', p. 327.

65. Mallon, D.P. 2003. *Gazella arabica*. In: IUCN 2006. *2006 IUCN Red List of Threatened Species*. <www.iucnredlist.org>. Downloaded on 14 December 2008.

66. Colin Groves, 'The Taxonomy of Arabian Gazelles', in Kushal Habibi, Abdulaziz H. Abuzinada and Iyad A. Nader (eds.), *The Gazelles of Arabia*, Riyadh 1997, p. 40.

67. 'Es ist aber unverzeihlich, daß man in Triest oder Berlin eine so kostspielige Reise so verhunzt [...] unsre beste Jugendzeit verliert sich im Wollen und Nichtkönnen.' Ehrenberg to Hemprich, 16 January 1824, Stresemann, *Hemprich und Ehrenberg*, p. 109; Ehrenberg to Lichtenstein, 3 April 1824, ibid., p. 113.

68. Hemprich to Ehrenberg, 15 January 1824, Stresemann, *Hemprich und Ehrenberg*, p. 111.

wrote to Hemprich that, during his friend's absence, he had already discovered forty new genera of animals on which he had seen 'Savigny and Co. stumble'. And he concluded: 'If there is anywhere to expect *Eclat* from our hard work, it is here at the Red Sea.'[69] But this did not convince Hemprich When still no news had come from Lichtenstein or from Brandenburg, the Prussian consul in Trieste, by 10 February, he called Ehrenberg back to Cairo.[70] Hemprich made it quite clear that he had decided to go back to Prussia at all costs and that he would go alone if Ehrenberg wanted to stay on. Would that Ehrenberg had heeded this advice, both for the sake of Hemprich and in the interest of zoology. As it was, he argued that they could hardly return home, unsummoned and with their work unfinished. He agreed that there were no means for another major expedition, but that they could continue work in Quseir, await the passing of the plague, and 'then, in God's name, set sail'.[71] How he convinced his friend is not recorded in the letters. Ehrenberg first went to Cairo to settle affairs and then joined Hemprich in Alexandria where he talked him over. Ehrenberg wrote to inform Lichtenstein of the change of plans. They proposed staying in Egypt during the plague and move to the Lebanon and Syria from April to August 1824 to study the flora of these countries. They would return via Alexandria to Quseir on the Red Sea, where they would stay for a couple of months, and finally proceed towards Jiddah, Mocha and Massawa in January and explore the coastal region there for most of 1825. They would thus return to Trieste in January 1826.

Awaiting orders and money from Europe—six months had passed since Hemprich had gone to Alexandria—at the beginning of April 1824 Ehrenberg's patience was at an end. He angrily demanded how it was that 36 ships had come from Trieste, all without letters or money? Why had the money which had been promised them in August 1823 still not been forwarded by Brandenburg? He bitterly concluded that only one thing was clear, namely, 'that our game has been spoiled for the third time, that we have been fooling around in Egypt for four years'.[72] Two weeks later news was brought to them that the Prussian consul in Trieste had gone bankrupt and committed suicide. At this point the naturalists decided to wait no longer and leave for the Lebanon.

The Lebanon

The short sea voyage to Beirut on board the Austrian brig *Il Marinayo* was tedious and not without discomfort, but once on the spot the naturalists were immediately at work again. They found that much had to be done on the flora and on reptiles. The bird species they found at first were mainly European ones. They explored the area between Beirut and Tripoli, crossed the valley of Beqa'a, and visited Baalbek. Few

69. 'Hier habe ich seit Deiner Abreise schon 40, sage vierzig neue Thiergattungen in die Zoologie gebracht, habe Savigny u. Consorten stolpern gesehen u. wenn irgendwo *Eclat* zu erwarten steht von der mühseligen Arbeit, so ist es hier am rothen Meere.' Stresemann, *Hemprich und Ehrenberg*, p. 108.

70. 'Laß uns, lieber Freund u. Unglückstgefährte, bald heimkehren, ehe Pest und andere politische Umstände erst uns die Abreise unmöglich machen.' Stresemann, *Hemprich und Ehrenberg*, p. 111.

71. Ehrenberg to Hemprich, 22 February 1824, Streseman, *Hemprich und Ehrenberg*, p. 112.

72. '[...] daß unser Spiel uns zum drittenmale verdorben, daß wir 4 Jahre lang in Egypten im Narrenkleide waren'. Ehrenberg to Lichtenstein, 3 April 1824, Stresemann, *Hemprich und Ehrenberg*, p. 113.

reports survive of this part of the expedition. From a letter to Lichtenstein written on 3 August it is clear that the naturalists felt slightly disappointed by their collections: they had found an interesting dog species, a weasel, three bat species, and about 10 new birds. There were also some fine lizards, and some snakes. Hemprich was bitten trying to catch a viper and would have been 'soon in the other world'. Luckily, his health was strong enough to survive the bite. He labelled the animal *Lachesis libanotica,* but never described or published it, and it must therefore be considered a *nomen nudum*. In 1898 it was given the name *Vipera bornmuelleri* Werner.[73] Today the species is classified as threatened with extinction, occurring only in a limited area in the Golan Heights.

'There's a bear in our crates,' Ehrenberg wrote to his father from Beirut. It was a specimen of the Syrian Brown Bear from Bischerre on Mount Makmel and was described and depicted as the very first species in *Symbolae physicae*.[74] Today the subspecies *Ursus arctos syriacus* Hemprich & Ehrenberg (1828) is known from a wide area extending from Asia Minor to the Caucasus and Iran, but it is no longer to be found in Lebanon, where it disappeared at the beginning of the twentieth century.

Although Ehrenberg and Hemprich felt initially disappointed with the bird species in Lebanon, many of these were later described by Ehrenberg in *Symbolae physicae* (1833) as subspecies of known specific entities. These include the Eastern Rufous Bushchat (*Cercotrichas galactotes syriaca*), a Yellow-billed Chough they caught in Beirut (*Pyrrhocorax graculus digitatus*), and the Levantine Black Redstart (*Phoenicurus ochruros semirufus*). Others have since been raised to specific rank, such as the White-spectacled Bulbul (*Pycnonotus xanthopygos*). Ehrenberg also described new bird species which still stand today, such as the Syrian Woodpecker (*Dendrocopos syriacus*) and a warbler which Hemprich had labelled as *Hippolais languida*. This last species is known as Upcher's Warbler. The English ornithologist Henry Baker Tristram found the description in *Symbolae physicae* rather vague, and re-described it thirty years later as *Hippolais upcheri*, after his friend Henry Morris Upcher. Ehrenberg's name has, of course, priority over Tristram's.

Apart from the new entities described by Ehrenberg in *Symbolae physicae*, there were also birds collected by Hemprich in the Lebanon which other naturalists would use to describe new species. For example, a great number of individual specimens of a bird labelled 'Accentor? bicornis' were sent to Berlin. When Ehrenberg did not use them in his publication of 1833, Lichtenstein sold one to ornithologist Christian Ludwig Brehm, who described it in 1842 as *Phileremos bicornis*.[75] The nuthatch *Sitta neumayer syriaca*, caught in Bischerre in July, was likewise described by the Dutch ornithologist Coenraad Temminck in 1835 using an individual specimen Lichtenstein had sold him.

73. In 1999 the species was placed under the genus *Montivipera* by G. Nilson, B. Tuniyev, C. Andrén, N. Orlov, U. Joger and H.-W. Herrmann, 'Taxonomic Position of the Vipera Xanthina Complex', *Kaupia*, 8, 1999, pp. 99–102.

74. The *Quarterly Review* did not fail to point out that this was the very species which devoured the 42 children who mocked Elisha in 2 Kings 2:23–4 (vol. 114, 1863, p. 61).

75. Now *Eremophila alpestris bicornis*. See Mary LeCroy, *Type Specimens of Birds in the American Museum of Natural History*, New York 2003, p. 29.

Arabia and the Yemen

NEXT OPENING
August-Friedrich Schmidt, hand-coloured lithograph of the Syrian Brown Bear (*Ursus arctos syriacus*) from Bischerre on Mount Makmel.
Christian Gottfried Ehrenberg, *Symbolae physicae seu Icones et descriptiones corporum naturalium novorum aut minus cognitorum quae ex itineribus Libyam Aegyptum Nubiam Dongalam Syriam Arabiam et Habessiniam* …, Pars Zoologica I, Berlin 1828, plate I
The Arcadian Library, 10503

'We have been condemned never to reach our destination,'[76] Ehrenberg commented at the outset of the last part of their expedition, when, in September 1824, they were back in Egypt to prepare for a second exploration of the Red Sea area. When money had finally arrived, their most important servants and assistants had abandoned them in Syria. One, a man called Trevi from Leghorn, who both acted as cook and pressed the plants they collected, had stolen some of their money and disappeared in the Syrian mountains, and the two Germans, Burkhardt and Heinrich Falkenstein, who had been with them for the last fifteen months, had decided not to join them on their trip south but to try their luck with Lady Hester Stanhope, the eccentric granddaughter of the elder Pitt who lived alone on a hilltop near Sidon in the Lebanon. The loss of these essential assistants made Ehrenberg doubt whether they should persevere with their plans or give up at this stage.[77] Besides, there were reports of political unrest in Upper Egypt.

At the end of September Burkhardt and Falkenstein returned to Egypt. Having met with no success with Lady Hester Stanhope, they had regretted their desertion. Burkhardt was dying of fever—upon which Hemprich drily remarked 'it serves him right'[78]—and he died in Damietta the following month. A muted Falkenstein let the two naturalists know he was willing to follow them wherever they wanted to go. The hunter Martin Bretzka too, who had been employed by Hemprich before, was willing to join them again. With an Arab as cook, and the painter Finzi who was to help Ehrenberg with the drawing of new species, prospects suddenly changed and the naturalists felt more hopeful about the last part of their expedition.

In scientific circles back in Germany reactions to the final part of Hemprich and Ehrenberg's expedition were less optimistic. The curator of the Royal Herbal in Berlin, Dietrich Franz Leonard von Schlechtendahl, recognized the importance of the continuing work in the Near East, but 'at the same time I feel uneasy about the fact that you are not returning, that you still want to penetrate the Abesha lands [Ethiopia/Eritrea], never satisfied, never resting.'[79] Carl Wilhelm Eysenhardt, professor of natural history and botany at the University of Königsberg, issued a direct warning to the two naturalists: 'If one of you should die, this would, apart from all human considerations, be a far greater loss for science than the failure to explore Abyssinia or anywhere else.'[80] But such warnings came much too late to have any influence on Hemprich and Ehrenberg's decision. By the beginning of December 1824 they had already moved to Suez and had arranged to sail on an Arab vessel to Jiddah. The expedition had started badly. At the outset the painter Finzi had

76. 'Wir sind verurtheilt nie zum Ziele zu kommen.' Ehrenberg to Lichtenstein, 21 September 1824, Stresemann, *Hemprich und Ehrenberg*, p. 123.
77. Ehrenberg to Lichtenstein, 21 September 1824, Stresemann, *Hemprich und Ehrenberg*, p. 124.
78. 'Die Lektion ist gut gewesen.' Hemprich to Ehrenberg, 29 September 1824, Stresemann, *Hemprich und Ehrenberg*, p. 128.
79. 'Zugleich wird es mir dabei etwas ängstlich, dass Ihr noch nicht rückkehret, wieder eindringen wollet gen Habesch, nimmer zufrieden, nimmer rastend und ruhend.' Schlechtendahl to Ehrenberg, 1 January 1825, Stresemann, *Hemprich und Ehrenberg*, p. 135.
80. 'Stirbt einer von Euch, so ist, abgesehen von allen andern menschlichen Rücksichten, der Verlust für die Wissenschaft, weit grösser, als durch Unterlassung einer Excursion in Abyssinien, oder sonst wo, hinein.' Eysenhardt to Hemprich and Ehrenberg, 19 November 1824, Stresemann, *Hemprich und Ehrenberg*, p. 135.

shot the foot of a Bedouin to pieces by accidentally firing his rifle. This caused much delay and the compensation they had to pay meant a considerable loss of money. In the end Finzi had to travel from Cairo to Suez alone long after the others had left in order not to expose them all to revenge.

After a short voyage they reached Jiddah on 28 December. Here they were forced to stay in the town, waiting for a ship to carry them to Abyssinia. There was little to do in Jiddah, as Christians were not allowed to cross the gates of the town lest they desecrate them, Ehrenberg explained to his father. According to Hemprich, on the other hand, they were not allowed to leave the town because the authorities were afraid the 'Christian dogs' would go and nose around in the holy city of Mecca. But notwithstanding these restrictions on moving in the area, they were able to do some interesting work on the coast. They finally collected specimens of most of the fish Forsskål had described, and caught interesting birds, including numerous individual specimens of an unknown 'sparrow'. In a letter to Lichtenstein Hemprich also mentions a new sandfish (a skink). The sparrow was eventually described by Rüppel as *Ploceus galbula* in 1837, while Wiegmann baptized the reptile *Scincus meccensis*. When Forsskål arrived in Jiddah in October 1762 he found the time of year was not propitious for botany. Things were clearly different in January. One day the naturalists managed to sneak out of Jiddah into the mountains towards Mecca and many interesting plants were found, some new, but many of them Forsskålean species. But these were not the reason why they had ventured south once more. Ehrenberg felt that he was merely following in Forsskål's footsteps, and was impatient to move on to unexplored regions: 'It should be possible to go on an excursion in Arabia and find something the Danes had not found.'[81] 'It is high time we left this place,' Ehrenberg wrote to his father.[82]

On 25 January the naturalists were ready to continue the voyage down the Red Sea to Massawa, which was to serve as their base for excursions into Abyssinia. They sailed close to the Arabian coast between coral reefs and islands towards Qunfudhah, collecting fish and other sea animals on the way, going on land for plants and animals when the vessel dropped anchor for the night. The first night they went on land and found such a rich desert flora that they decided to stop there for a day. The delay was well justified, as they managed to collect a great number of flowering plants. During an excursion inland from Mersa Ibrahim a few days later they found Forsskål's *Salicornia perfoliata*.[83] On 6 February they reached Qunfudhah, from where they had planned to explore the Asir mountains. Hemprich first went on an excursion towards the north, but the camel drivers refused to venture into the higher mountains, and he came back with little of interest. A week later it was Ehrenberg's turn. He moved east with their three European hunters and four Turkish soldiers. Again Ehrenberg found many of Forsskål's species. On the mountain tops baboons (*Papio hamadryas*) could be seen regularly, but attempts to shoot some individuals came to nothing. In his diary Ehrenberg recounts how one endeavour to ambush a small group of

81. 'Irgend wo werden wir ja ankommen können zu einer Exkursion in Arabien u. werden schon noch allerlei finden, was die Dänen dort ließen.' Ehrenberg to Link, 22 January 1825, Stresemann, *Hemprich und Ehrenberg*, p. 140.

82. Ehrenberg to his father, 1 January 1826, Stresemann, *Hemprich und Ehrenberg*, p. 138.

83. See above, pp. 90–1.

animals ended in a spectacular failure that provoked the laughter of the on-looking Arabs. As Ehrenberg and his three hunters cautiously circled a mountain top where they had seen a mother with its young, Bretzka ventured a shot at the adult baboon. When Ehrenberg looked up he saw the animal standing upright in the act of wiping its flaming red backside with a handful of sand. The shot had got only a fourth of the way, and the animals retreated to safety with ease.[84] In *Symbolae physicae* Ehrenberg was the first to identify this species definitively with the sacred baboon, the attendant of Thoth of ancient Egypt.

They had more success with the hares in the desert, which were rather smaller and darker than the species they had caught in the Sinai, and considerably smaller than the common hare. Ehrenberg described it as the new species *Lepus arabicus*, but today it is considered a mere subspecies of the common hare. The animal is depicted in one of the most beautiful plates of *Symbolae physicae*.

Heavy rains and flooding made it inadvisable to stay in the area any longer, and rather than risking their collection Ehrenberg decided to return to Qunfudhah. On 4 March they set sail again. Anchoring near the volcanic isle of Ketumbul, they quickly explored the barren black crater and found interesting plant species (amongst which the new genus *Ketumbulia*) and beautiful butterflies. On 6 March they reached the Farasan Islands, where they hunted the famous *Gazella arabica* (see above). Giza, on the coast, rendered 'a treasure of new fish species'. At the end of the month they reached the Yemen, and Hemprich lost no time in retracing Forsskål's steps from Luhayyah to Mawr, while Ehrenberg prepared their new finds for shipment to Berlin. Animosities between the villages in the hills were reported, and Ehrenberg sent a messenger to Hemprich telling him to return to the coast. Hemprich returned just in time: the village he had planned to visit was ransacked and many villagers, including women, were ruthlessly murdered. As nothing further could be done in the area, and because Forsskål had already done significant work there,[85] the naturalists crossed the Red Sea for their last destination. On the way they cast anchor at the Dahlak islands,[86] where they found a dead dolphin washed ashore, and collected its skull. It turned out to be a specimen of the Red Sea Bottlenose Dolphin (or, Indian Ocean Bottlenose Dolphin), then a still unknown species. Ehrenberg described it in 1833 as *Tursiops aduncus*.

NEXT OPENING
Friedrich Bürde, hand-coloured lithograph of hares from Dongola.
Symbolae physicae seu Icones et descriptiones corporum naturalium novorum aut minus cognitorum quae ex itineribus Libyam Aegyptum Nubiam Dongalam Syriam Arabiam et Habessiniam …, Pars Zoologica I, Berlin 1828, plate XIII
The Arcadian Library, 10503

FOLLOWING OPENING
Friedrich Bürde, hand-coloured lithograph of the Arabian Leopard (*Panthera pardus nimr* Hemprich & Ehrenberg).
Christian Gottfried Ehrenberg, *Symbolae physicae seu Icones et descriptiones corporum naturalium novorum aut minus cognitorum quae ex itineribus Libyam Aegyptum Nubiam Dongalam Syriam Arabiam et Habessiniam …*, Pars Zoologica I, Berlin 1828, plate XVII
The Arcadian Library, 10503

Abyssinia

'Abyssinia lies before us so peacefully and invitingly that we have already forgotten the long and dangerous journey it took us to get here,' Ehrenberg wrote to Lichtenstein as soon as they reached the islands of Massawa. In retrospect these words have an ironic sound. Hemprich had lost no time in exploring nearby Hargigo as well as the islands of the Dahlak Archipelago off the Eritrean coast, while Ehrenberg remained behind with Finzi to sort out their collections and make a complete series

84. Stresemann, *Hemprich und Ehrenberg*, p. 157.
85. Moreover, Ehrenberg writes that they had by then found most of Forsskål's plants (to Lichtensteuin, 30 April 1825, Stresemann, *Hemprich und Ehrenberg*, p. 142).
86. The name Ehrenberg recorded for the island where they found the dolphin is 'Belhosse'.

of pictures of fish. 'I have been suffering from a bit of rheumatic fever for three days,' Hemprich soon admitted, 'and it makes me feel uncomfortable.'[87] A wry warning.

The excursion to Hargigo and the Dahlak Islands was meant as a preliminary trip to explore the region. Hemprich would next go into the interior as far as Adwa (Ethiopia) for a longer period to collect the bigger animals (elephants, lions and rhinos),[88] while, once Hemprich had returned, Ehrenberg would venture inland a second time to collect plants and smaller animals. The first news from Hemprich was promising. Near Hargigo they finally managed to obtain a baboon, a fine adult specimen. Hemprich also reported numerous unknown birds and interesting forms of jackal, boar, gazelle and hare. Amongst the new Abyssinian birds were *Nectarinia habessinica*, the Rosy-patched Bushshrike (*Rhodophonus cruentus*), and a woodpecker and a hornbill, both of which Ehrenberg dedicated in 1833 to his friend: *Dendropicos fuscescens hemprichii* and *Tockus hemprichii*. The jackal and hare have proved of dubious status and can at best be seen as a subspecies of the common jackal and the common hare. The first was not even depicted in *Symbolae physicae*. The boar was described by Ehrenberg in 1828, but Cretzschmar had previously used Hemprich's specimen for a description of the animal in 1825. Ehrenberg decided to dedicate the gazelle to Hemprich by calling it *Antilope hemprichiana*, but again the species had previously been published in 1816 by Desmarest. In 1837 it was placed by Ogilby in a new genus, following the Abyssinian name 'madoqua' Henry Salt had picked up for a skin he had deposited in the British Museum in 1811[89] and which Hemprich had given to his specimen.

When Hemprich joined Ehrenberg again in Massawa at the beginning of June, he fell victim to the epidemic which was raging on the island. He was soon laid low with a '*febris intermittens perniciosa sudorifera*'. The hunters and servants followed, and, Ehrenberg, who had had the fever twice himself during Hemprich's absence, feared they were fast running out of medicine, quinine and money. Despite the fact that Hemprich's pulse soon became somewhat more regular and his appetite returned, Ehrenberg, who had expected so much from this last part of their expedition,[90] had a fit of despair: 'What has become of all our hopes!' He wrote to Lichtenstein, 'We have come to its very door and now we cannot enter!'[91] Ever since the beginning of the expedition Ehrenberg had repeatedly written home about the dangers, hardships and disappointments that befell them, but in the end their problems were always overcome. Not this time. Bretzka felt so ill that he had to leave them, and Heinrich Niemeyer, one of the hunters, died on 15 July. And Ehrenberg's next letter to Lichtenstein, written in Cairo, bore the following news: 'The dice have been

Friedrich Bürde, hand-coloured lithograph of wild goats (*Capra ibex nubiana*). Christian Gottfried Ehrenberg, *Symbolae physicae seu Icones et descriptiones corporum naturalium novorum aut minus cognitorum quae ex itineribus Libyam Aegyptum Nubiam Dongalam Syriam Arabiam et Habessiniam* ..., Pars Zoologica I, Berlin 1828, plate XVIII
The Arcadian Library, 10503

NEXT OPENING
Friedrich Bürde, hand-coloured lithograph of *Papio hamadryas*. Christian Gottfried Ehrenberg, *Symbolae physicae seu Icones et descriptiones corporum naturalium novorum aut minus cognitorum quae ex itineribus Libyam Aegyptum Nubiam Dongalam Syriam Arabiam et Habessiniam* ..., Pars Zoologica I, Berlin 1828, plate XI
The Arcadian Library, 10503

87. 'Ich bin seit 3 Tagen von einem rheumatischen Fieberchen befallen, was mich etwas ärgerlich macht.' Hemprich to Ehrenberg, [10] May 1825; Stresemann, *Hemprich und Ehrenberg*, p. 143.

88. Ehrenberg to Lichtenstein, 30 April 1825, Stresemann, *Hemprich und Ehrenberg*, p. 142.

89. '[...] another very small species, little bigger than a hare, found commonly also at the Cape and Mosambique, [is] called in Tigré 'madoqua', Henry Salt, *A Voyage to Abyssinia, and Travels into the Interior of that Country, Executed under the Orders of the British Government, in the Years 1809 and 1810: In Which Are Included, An Account of the Portuguese Settlements on the East Coast of Africa, Visited in the Course of the Voyage; A Concise Narrative of Late Events in Arabia Felix*, Philadelphia and Boston 1816, p. 428.

90. 'Aus Abyssinien erwarten wir viel, sehr viel.' Ehrenberg to Lichtenstein, 13 May 1825, Stresemann, *Hemprich und Ehrenberg*, p. 144.

91. 'Wo sind jetz unsre Hoffnungen! ... Wir kommen bis an die Thür und nicht hinein!' 2 July 1825, Stresemann, *Hemprich und Ehrenberg*, p. 145.

cast! Our last hopes have been destroyed, dreadfully destroyed! Hemprich died in my arms on 30 June.'[92] There are no accounts of how Hemprich died, but judging from a death from Massawa fever which Salt witnessed in 1810, it must have been atrocious:

> [...] *they found him still alive, though suffering under the violent delirium which commonly attends the last stage of a putrid fever. He had been most injudiciously treated, and was chained, with his face downwards on a couch, so that his body was bruised, and his skull almost fractured, by the vain efforts he had made to release himself* [...] *These are the fevers which so often attack strangers who come down from the interior, and which produce in the minds of the Abyssinians that great dread and horror of the coast which they generally entertain.*[93]

To save their collection, and, above all, to save his own life, Ehrenberg hurried away from Massawa. Almost overcome with fatigue and illness—he too had contracted the fever once more—he crossed the Red Sea in an open vessel, getting terribly wet during a three-day storm, and reached Jiddah 'more dead than alive'. He despaired of ever reaching Egypt. During the crossing to Quseir five passengers died, and he arrived in Cairo suffering from acute dysentery. When, after a long voyage, Ehrenberg finally arrived in Trieste, he was a broken man. 'All the people take me for an old man,' he wrote to Camisso, and in a letter to Nees von Esenbeck he wondered: 'Are all the results of this journey worth 9 human lives, or even only one?? I set out light-hearted, in health and happy, but have come back in bad health, sad and gloomy.'[94]

92. 'Das Loos ist gefallen! Die lezte unsrer Hoffnungen zertrümmert, grässlich zertrümmert! Hemprich starb am 30sten Juny in meinem Arme.' Stresemann, *Hemprich und Ehrenberg*, p. 145.

93. Henry Salt, *A Voyage to Abyssinia*, pp. 168–9.

94. 'Sind wohl die Resultate dieser Reise 9 Menschenleben, oder auch nur das Eine werth?? Ich ging leichtsinnig, gesund und froh von Ihnen und komme krank, betrübt und schwermüthig wieder.' Ehrenberg to Chamisso, 6 December 1825, Stresemann, *Hemprich und Ehrenberg*, p. 146; to Nees von Esenbeck, 10 December 1825, ibid., p. 147.

CHAPTER 4

BALFOUR, SCHWEINFURTH AND THE SOQOTRA EXPEDITIONS OF 1880–1881

In 1878 the British Association for the Advancement of Science (BAAS) decreed that £100 was to be placed at the disposal of a committee to set up a scientific expedition for the investigation of the natural history of the island of Soqotra. Now part of the Yemen, but situated off the coast of Somalia in the Indian Ocean about 500 miles south-east of Aden, the island had long been an important stop on the trade route from Aden to India. The ancient Greeks settled there in 330 B.C., and St Thomas the Apostle is reported to have been shipwrecked there and to have introduced Christianity. In the Middle Ages Soqotra became famed for its 'dragon blood', a red resin used as a coagulant in the treatment of wounds, and Marco Polo did not fail to visit it. In the second book of his *Travels* (chapter 32), he mentioned the island's trade customs, its Christianity, its piracy and its sorcerers. In the early sixteenth century the Portuguese invaded the island, but had withdrawn by the middle of the century. In 1608 the first ship of the British East India Company visited Soqotra, but it was not until 1834 that it was occupied by its forces, who stayed there for five years.

The decision of the BAAS to investigate the flora and fauna of Soqotra came soon after Britain had signed a peace treaty with the sultan of Soqotra in 1876. The Soqotra committee decided that the best time to visit the island was from November to March and appointed Colonel H.H. Godwin-Austen, a competent naturalist, to lead the expedition in 1879. It was estimated that the complete cost of the expedition would amount to £300. Godwin-Austen, however, was no longer available when the autumn of 1879 arrived, and, on 20 December, Joseph Hooker, member of the committee, wrote to Isaac Bayley Balfour, a young professor of botany at the University of Glasgow. Balfour was a good choice. He was the son of the botanist John Hutton Balfour, professor of botany at the University of Edinburgh, where he graduated in science in 1873. He had also studied at the universities of Würzburg and Strasbourg, and taken part in a scientific expedition to Rodrigues. In 1875 he obtained a PhD from the University of Edinburgh for his work on the botany of this Mascarene island. Balfour was also 'nothing loath to be the first to unravel some of the mystery which for so long had clung around Socotra',[1] but he explained to Hooker that he had to be back in Glasgow by April and that he could hardly hope to reach Soqotra before the beginning of February. However, he added, if the BAAS would accept a mere two months' work on the island, he was willing to go. Joseph Hooker answered that he felt certain this would be sufficient for obtaining knowledge of the natural history of the place.

Although Soqotra had long been an important stop for travellers to India, no systematic botanical or zoological research had been carried out on it. A number of naturalists had done a bit of collecting in passing and several species were collected by the officers of the East India Company forces during the occupation of the island from 1834 to 1839. Some of these were sent to Kew Gardens, while others were

1. Isaac Bayley Balfour, *Botany of Socotra,* Edinburgh and London 1888, p. x.

described in a memoir by Lieutenant James Raymond Wellsted.[2] But the importance of Balfour's expedition lay in the fact that it was the first large-scale attempt to investigate the natural history of Soqotra.

After very little preparation Balfour travelled to Marseilles on 9 January and boarded a French steamer bound for Aden, which he reached two weeks later. Here Balfour and Alexander Scott (a botanical collector and gardener at the Royal Botanic Garden of Edinburgh) were joined by Lieutenant Charles J. Cockburn, an able draftsman, and he hired native servants and an interpreter. On 6 February H.M.S. *Seagull* conveyed them to Gollonsir Bay north-west of Soqotra, and letters of recommendation from Aden were sent to the sultan in Hadibo. Deeming it unsafe to land on the coast without an official permit, they had to wait ten days before an invitation from the sultan arrived.

Collecting specimens on the island was not without difficulties. First, the time they had at their disposal was hardly sufficient for the task they had set themselves, and as the purpose of the expedition was to obtain a representative collection of the fauna and flora, Balfour tried to cover as much ground as possible, which meant that 'much barren land was travelled over, and many rich and fertile spots were necessarily only superficially looked at'.[3] The members of the expedition also discovered that they could by no means do as much work as they desired because of the intense heat during that time of the year. Moreover, their inadequate knowledge of the geography of the island made it extremely difficult to find their way about: 'The chart based on Wellsted's observations is the only available one,' Balfour grumbled, 'and that is so incomplete and incorrect as to be almost useless to any one moving about the island.'[4]

Notwithstanding these setbacks, the expedition was a success. During their 44 days[5] on Soqotra Balfour collected over 500 dried specimens of flowering plants, while a few interesting items, such as a Dragon's Blood Tree, were dug out to be transported live to England. Although there was a disappointing 'paucity' of mammal species, birds, lizards, fish and land molluscs were plentiful, and 'the whole island teems with insect life.'[6] As for numbers, they collected 36 species of birds, 48 species of molluscs, 6 species of lizards, and 24 species of butterflies. Many of these were new, some of them endemic to Soqotra. There were few species of snakes, but apart from a viper which is also found on the Arabian peninsula, Balfour found that the three other snakes were peculiar to Soqotra, and he surmised that two of these constituted not just new species but new genera. Moreover, they amassed 500 specimens of rocks and minerals.

When Balfour returned to Glasgow in April 1880 he did not know that Emil Riebeck, as part of a major four-year expedition to the Far East to collect artefacts of indigenous populations, intended to stop on Soqotra and explore the island's mineral resources as well as its flora and fauna.

2. 'Memoir on the Island of Socotra', *Journal of the Royal Geographical Society*, 5, 1835, pp. 129–229.

3. Balfour, *Botany of Socotra*, p. xiv.

4. Balfour, *Botany of Socotra*, p. xiv.

5. Balfour mentions that they stayed for 48 days on Soqotra (Balfour, *Botany of Socotra,* p. xv); other evidence in Balfour's account, however, indicates that the expedition landed on the island on 16 February and sailed for Aden on 30 March.

6. Balfour, *Botany of Socotra*, p. xxx.

Riebeck was the son of the mining magnate and industrialist Carl Adolf Riebeck. He was born in 1853 and studied natural sciences at the universities of Karlsruhe and Leipzig. In 1880 he obtained a doctorate at the University of Freiburg with his inaugural dissertation *Beiträge zur Kenntniss des Pyropissits*. Soon after this he decided to travel, and he visited the Caucasus, Turkey and Syria in the company of Friedrich Mook and Carl Rosset. The expedition next explored Palestine, where they were taken hostage by a group of Bedouin and ransomed for money. From Cairo they explored the desert to the east, and then continued to Soqotra. From there they crossed to India, and ultimately got as far as Burma, Singapore, Java, China and Japan. The four-year expedition considerably damaged Riebeck's health, and he died in 1884, shortly after his return to Europe. His extensive ethnographical collections were donated to the Kunsthistorisches Museum in Berlin. Today his name is mainly linked to Riebeckite, a dark blue mineral which is mined industrially for its asbestos content, while a form which is pseudomorphed by quartz is used as a semi-precious stone known as a tiger's eye. The type locality of Riebeckite is Soqotra, and the mineral was named after the German explorer in 1888, four years after his death.

The Riebeck expedition set out as a three-man expedition, led, and sponsored, by Riebeck himself. Ten years older than Riebeck, Friedrich Mook brought some travelling experience to the project. He had studied theology at the University of Tübingen, where he was awarded a doctorate, and continued theological studies in Utrecht, before being ordained. However, he soon tired of clerical duties, and enrolled at the University of Berlin, and later at the University of Heidelberg, in order to study medicine. In 1874 he travelled for the first time to Africa and the Near East. Upon his return he wrote a critical work on Paracelsus, but after 1875 he travelled almost annually to the East to study the Egyptian Stone Age. In 1880 the results of these travels led to his best-known archaeological study: *Aegyptens vormetallische Zeit*. In the same year he decided to join Riebeck.

Little is known about the third member of the 'Riebeck'sche Expedition', Carl W. Rosset. He was a steadfast travelling companion to Riebeck on his four-year expedition to the Far East, and, after their return in 1884, was sponsored by Riebeck to explore the Maldives, about which he claimed that he was 'the first [European] who has undertaken a systematic exploration of the groups, and who for that purpose has taken up his abode among and associated with the people'. His observations were published as a finely illustrated travelogue.[7] In the late 1880s he travelled once more to India. His interests were principally those of an ethnologist, and he brought home an important collection of artefacts.

As the Riebeck expedition proceeded to Cairo to prepare for Soqotra at the beginning of 1881, two other Germans joined it. One was the experienced traveller Georg Schweinfurth, and the other a Dr Otto Mantey. Hardly anything is known about Mantey,[8] except that he was a physician residing in the Egyptian capital. Schweinfurth, on the other hand, is famous in the history of scientific exploration. Born into a prosperous German-Russian family, he dedicated his life to the exploration of Africa and Arabia. After studying botany and palaeontology at the universities of Heidelberg, Munich, and Berlin, he travelled to Egypt and the Sudan

7. 'The Maldive Islands', in *The Graphic*, 16 October 1886.
8. His name is alternatively spelled 'Mantay'.

The Riebeck Expedition diary, bound in soft roan.
Expedition Riebeck / Halle a.d. Saale (Allemagne). / Tagebuch 1881. / Dr Mantey, unpublished MS
The Arcadian Library, 16521

during the mid-1860s, while from 1868 to 1871 he travelled for the Humboldt Stiftung in Berlin up the Nile from Khartoum into the very heart of Africa, to the Bongo and Dinka peoples in south-western and southern Sudan, and as far south as the Uele River, which he discovered, in the north of the modern Democratic Republic of Congo. He later described these travels in *Im Herzen von Afrika* (1884), a book which was to prove a classic account of the exploration of the African interior. In 1872 he founded the Egyptian Geographical Society in Cairo and the following year he explored the Libyan desert. Having settled in Cairo during the second half of the decade, he led various excursions in Egypt and the Red Sea area during the years 1878–1880. In 1881 he was asked to lead the Riebeck expedition to the island of Soqotra.

If, apart from a handful of scientific publications, there is little published information about the English expedition to Soqotra, there is even less about the German one. But, while there are no extant diaries for Balfour's stay on the island, a handwritten account of part of Riebeck's expedition has recently come to light. This travel diary is now in The Arcadian Library. It is a day-to-day report labelled 'Expedition Riebeck / Halle a.d. Saale (Allemagne). / Tagebuch 1881. / Dr Mantey' and has printed page-numbering running from 2143 to 2380.[9] Despite its label, the diary covers the period from 13 November 1880 to 11 June 1881, and is written in at least three different hands.

The first part relates in short pencilled notes the expedition's excursions round the Dead Sea from 13 November to 13 December 1880. It gives a great deal of geographical information, where drinking water was to be found, day and night temperatures, the weather, encounters with the local population and the constant requests for bakshish. The entry for 26 November, for example, records how the members of the expedition excited the Arabs of the area who requested 30 francs when they decided to leave the site which had been appointed for their tents. They were surrounded by shouting armed men, who took hold of their horses, and they could not but pay. Every now and then there is mention of the vegetation in the diary, but only very generic information about animals and plants—generally no more than a reference to the presence of butterflies, crabs or fish.[10] The first part of the diary contains a few rudimentary sketches of an archaeological site, frequent references to Rosset, who had contracted a persistent fever while travelling around the Dead Sea, while Riebeck is only mentioned in passing. This part was probably written by Friedrich Mook.

The account of the expedition is rudely interrupted by a horizontal brace, underneath which one can vaguely discern the words 'Mook Tod', followed by a description, written in a different hand, of how, on 13 December 1880, Mook died near Jericho. As the water of the Jordan was too high for their horses to swim with their loads, Mook, who tried to shuttle the luggage across on ropes, remained stuck in a raft and drowned.[11] It is clear that this account must have been written by either Rosset or Riebeck.

The third part of the diary, which describes the trip from Cairo to Soqotra, is

9. Pages 2299 to 2348 are left blank.
10. *Tagebuch*, pp. 2153–4.
11. *Tagebuch*, p. 2198.

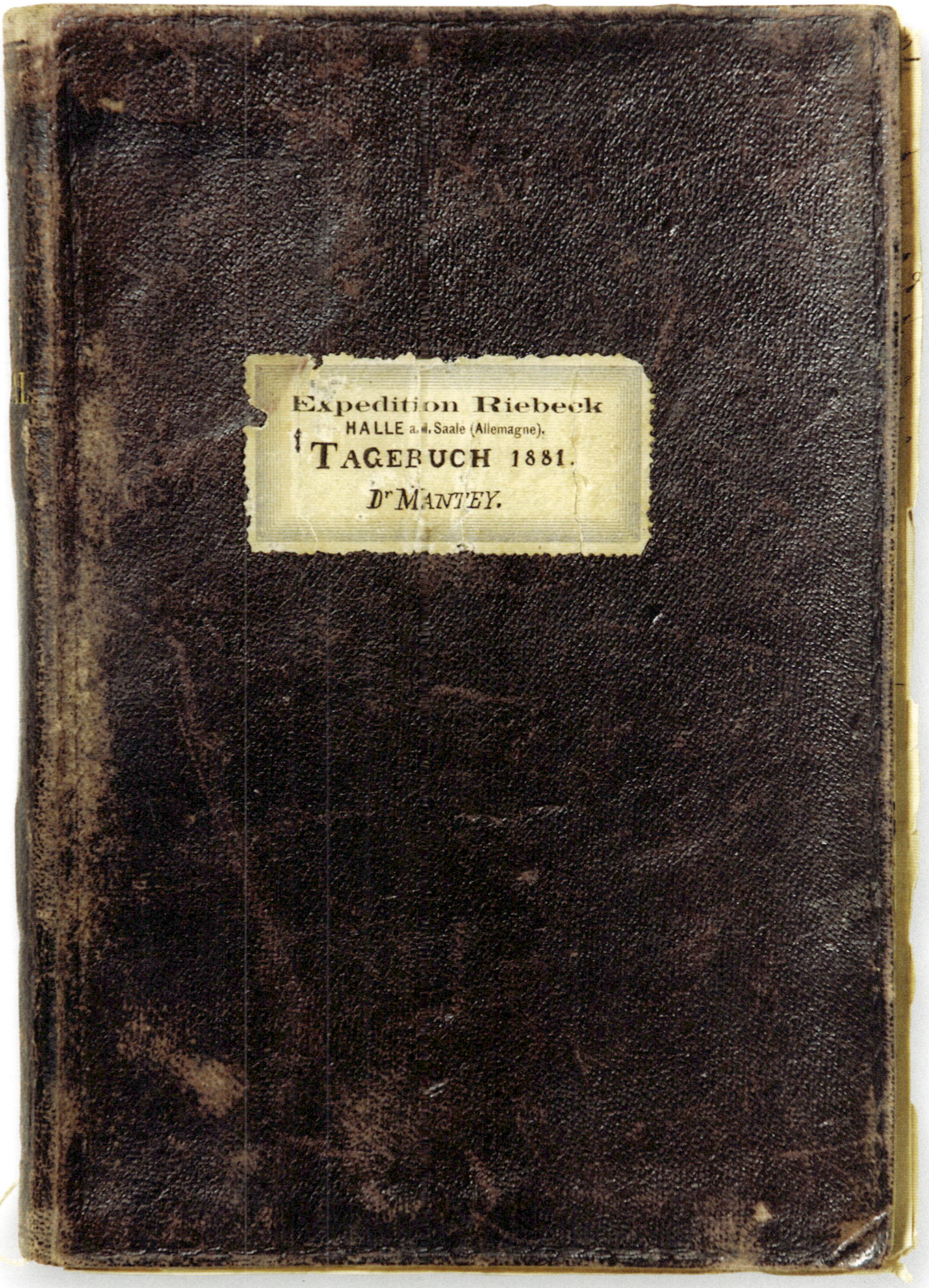
Expedition Riebeck
HALLE a. d. Saale (Allemagne).
TAGEBUCH 1881.
Dr MANTEY.

Otto Mantey, pen-and-ink drawing of Umbrella Thorn near Shuqrah.
Expedition Riebeck / Halle a.d. Saale (Allemagne). / Tagebuch 1881. / Dr Mantey, unpublished MS, pp. 2203–4
The Arcadian Library, 16521

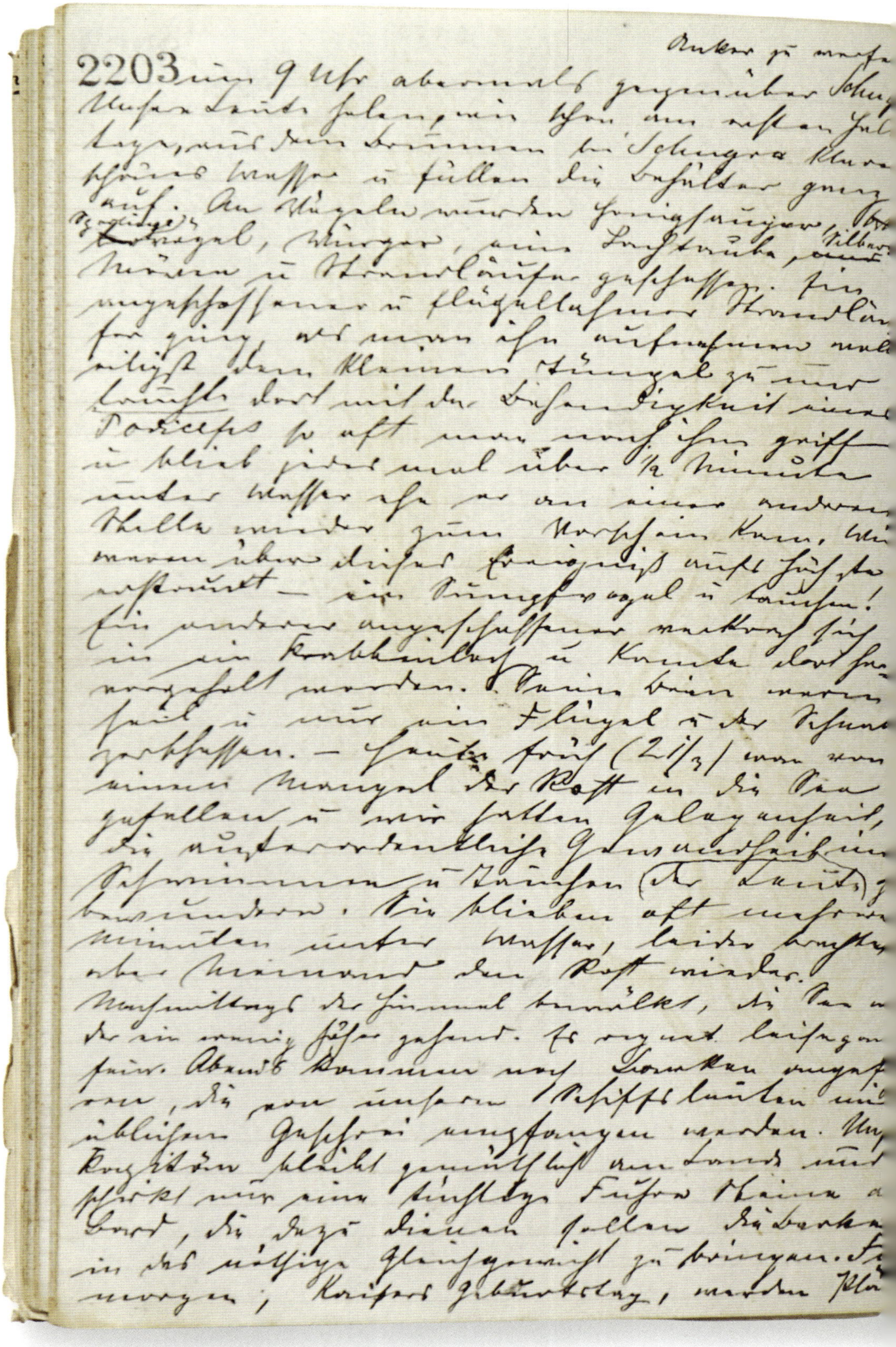

Acacia
spirocarpa.

Otto Mantey, hand-coloured drawing of jellyfish (*Rhizostoma*).
Expedition Riebeck / Halle a.d. Saale (Allemagne). / Tagebuch 1881. / Dr Mantey, unpublished MS, pp. 2209–10
The Arcadian Library, 16521

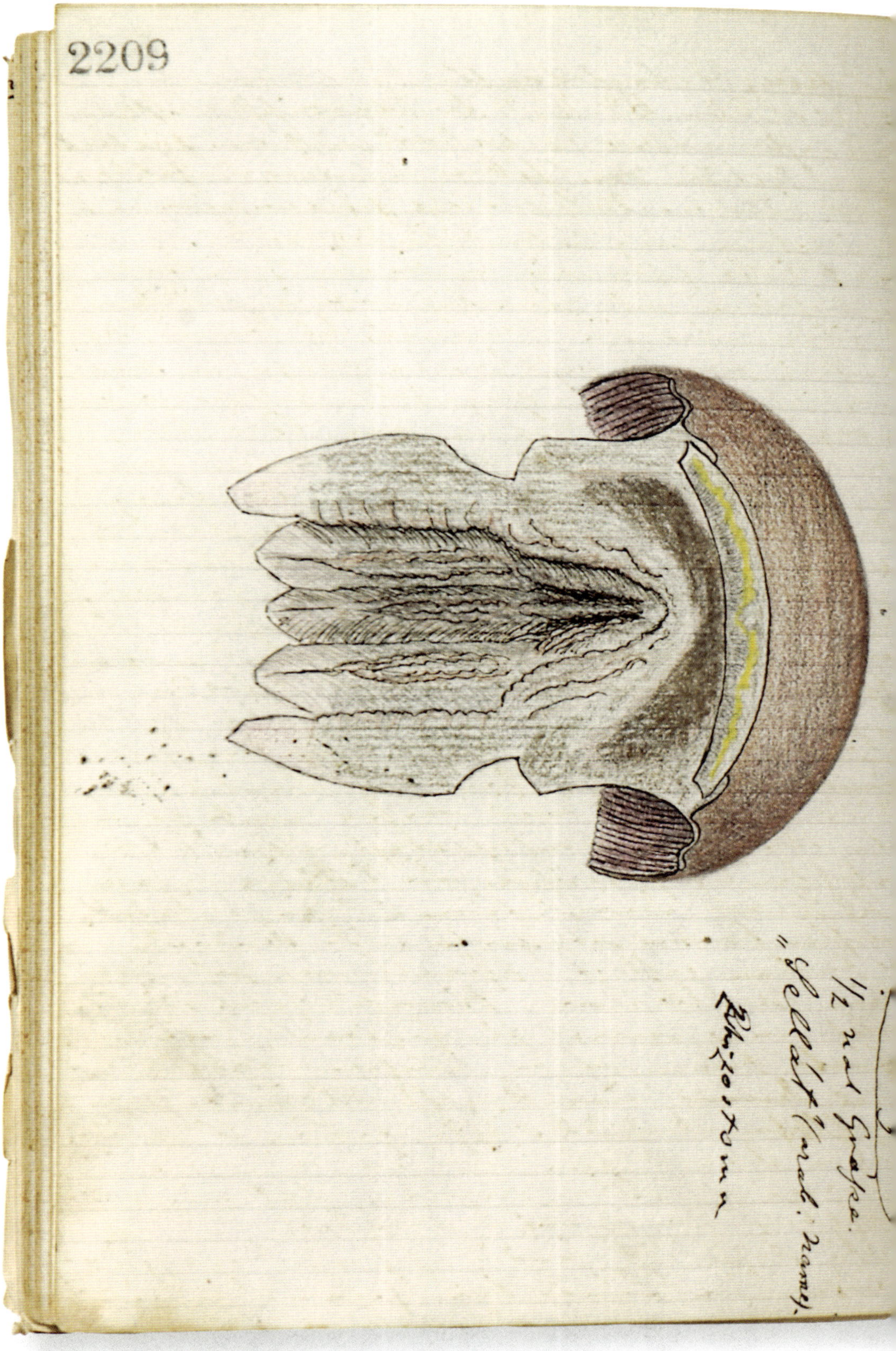

24 Maerz Donnerstag.
5h30 SO. 6h N.N.O. See viel ruhiger als gestern. Um 10 Uhr wird noch ein zweites Segel gesetzt. Wir halten immer denselben Curs inne. Nachmittags schwimmen viele Delphine ganz in der Nähe des Landes. Um ½ 5 Uhr sehen wir wieder Land; die Leute unterhalten sich ziemlich eifrig mit dem Engländer über die Namen, des kennen des Landes wegen sichtbaren Berge an der arabischen Küste. Angeblich sollen wir morgen in den Hafen von Bir Ali, der einzigen Bucht außer Aden an der Südküste Arabiens, einlaufen. Gegen 6 Uhr wurde die Segel fester beigesetzt und der Curs auf NO eingehalten. Die See ist so ruhig geworden, daß wir seit drei Tagen das erste mal wieder unseren Tisch im „Salon" aufschlagen können. 8h Abends Curs O.

25 Maerz Freitag. 5h30 N:NO. ruhige See. Himmel ziemlich klar. 8h O. Seit 6h30 Land zu sehen. 9h30 N NO. Es wird mehr und mehr gewendet. Wir sehen Nachmittags zwei Wale wahrscheinlich Pottfische. Die Leute erzählen sich von denen einer in den wundervollsten Farben irisirte. Die Änderung der Farben geht so schnell vor sich, daß kein Maler im Stande sein dürfte sie nur eine zu fixiren. Zwei Fischerboote kamen an unsere Barke und denen eine Anzahl kleiner sehr wohlschmeckender Fische gekauft wurden. Gegen Abend schwimmen wieder eine Menge Quallen derselben Art, wie vorgestern am Boote vorüber.
Die Küste ist ständig höhere Berge sind tiefer im Lande zu sehen. Einzelne

Otto Mantey, pen-and-ink drawing of Maidaha. *Expedition Riebeck / Halle a.d. Saale (Allemagne). / Tagebuch 1881. / Dr Mantey*, unpublished MS, pp. 2241–2
The Arcadian Library, 16521

den Samen nennen die Leute hier
Rabbisch und brauchen ihn zur Tabakskultur.
Soá-amer heißt hier die Fledermaus. (mit ain geschrieben)
9 Maerz. Dienstag. Ganz früh gegen 3 Uhr geht
es wieder langsam weiter. Fast Windstille.
Allmählich wird der Wind stärker & wir kreu-
zen unter sehr spitzem Winkel. Gegen
Mittag schwankt das Boot wieder so stark
daß vom Sich aufstellen keine Rede ist.
Um 1½ Uhr wird ungefähr 1 Kilometer
vom Lande Anker geworfen angesichts
Ras el Kelb. Einer der Bootsleute antwor-
tet auf die Frage, „warum?": „ras el Kelb –
ras el Kelb! – el hawwa balal." (das Hunde-
vorgebirge ist eben das Hundevorgebirge! der
Wind ist schlecht!) Doch wenigstens ein Trost.
Es scheint nur so, als wenn aus den an-
fangs angenommenen 14 Tagen 4 Wochen
werden könnten.
Abends beruhigt sich die See ein wenig.
Das Fortkommen scheint besonders durch
die hier sehr starke Strömung ~~erschwert~~ behindert
~~...~~ zu sein, die stündlich 2 Meilen
nach O u W betragen soll. Die Araber über-
treiben natürlich wieder & meinen bei
dem Wind könnten wir zehn Tage kreu-
zen & würden doch nicht um Ras el Kelb
kommen. Gegen Abend kommen ei-
nige Fischerbarken, denen sehr schmack-
hafte Fische entnommen wurden.
Die Küste ist sandig, im Hintergrund
immer höher aufsteigende Berge. Ein
kleiner Ort auf Ras el Kelb zu-
nächst ist zu sehen.
Abends leuchtet das ganze ~~...~~ Meer
in wunderbarer Pracht, einzelne
Funken (die Kämme der Wellen) leuchten
besonders stark.

Otto Mantey, pen-and-ink drawings tracing the skyline between Ras al-Kelb and Mukalla.
Expedition Riebeck / Halle a.d. Saale (Allemagne). / Tagebuch 1881. / Dr Mantey, unpublished MS, pp. 2243–50
The Arcadian Library, 16521

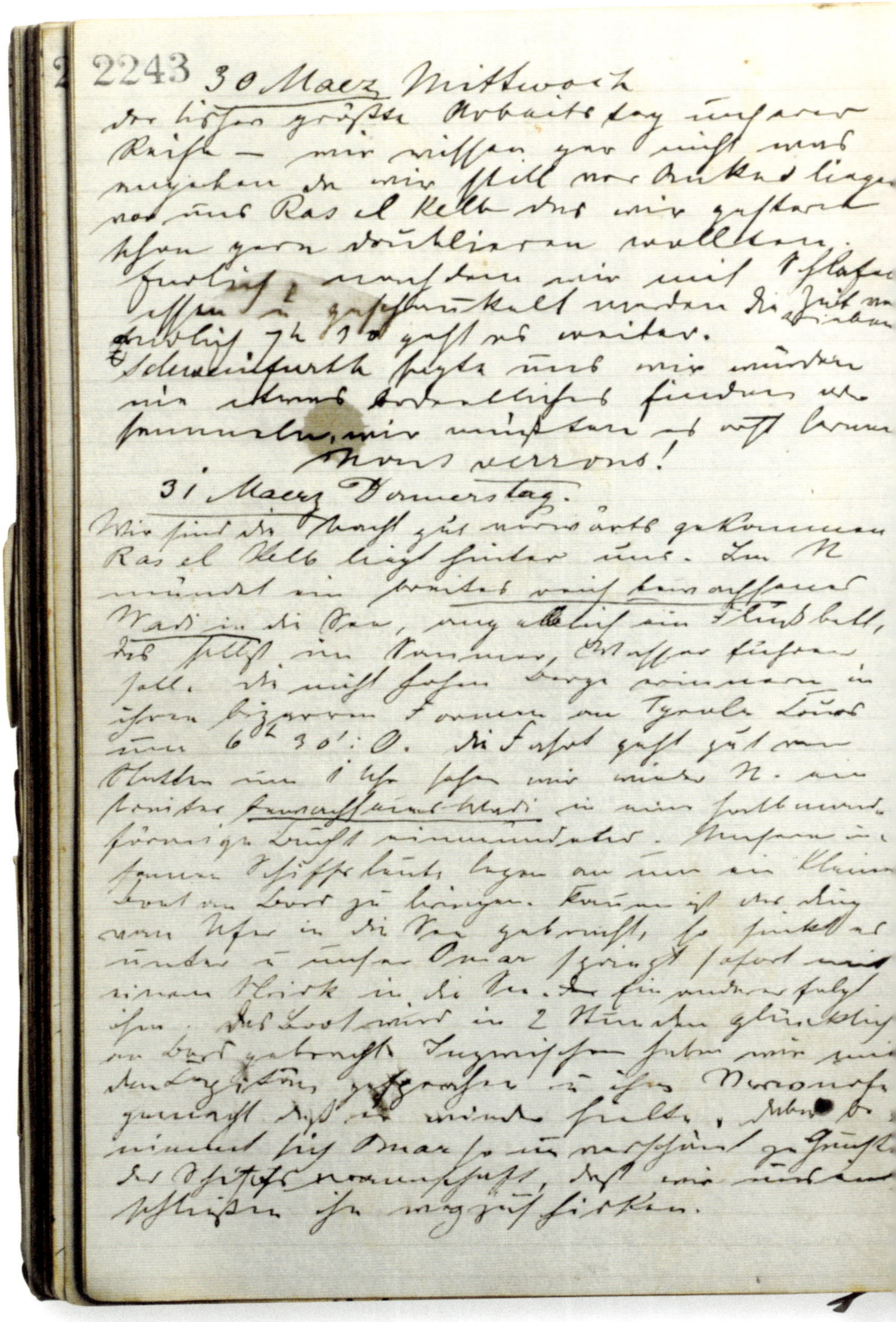

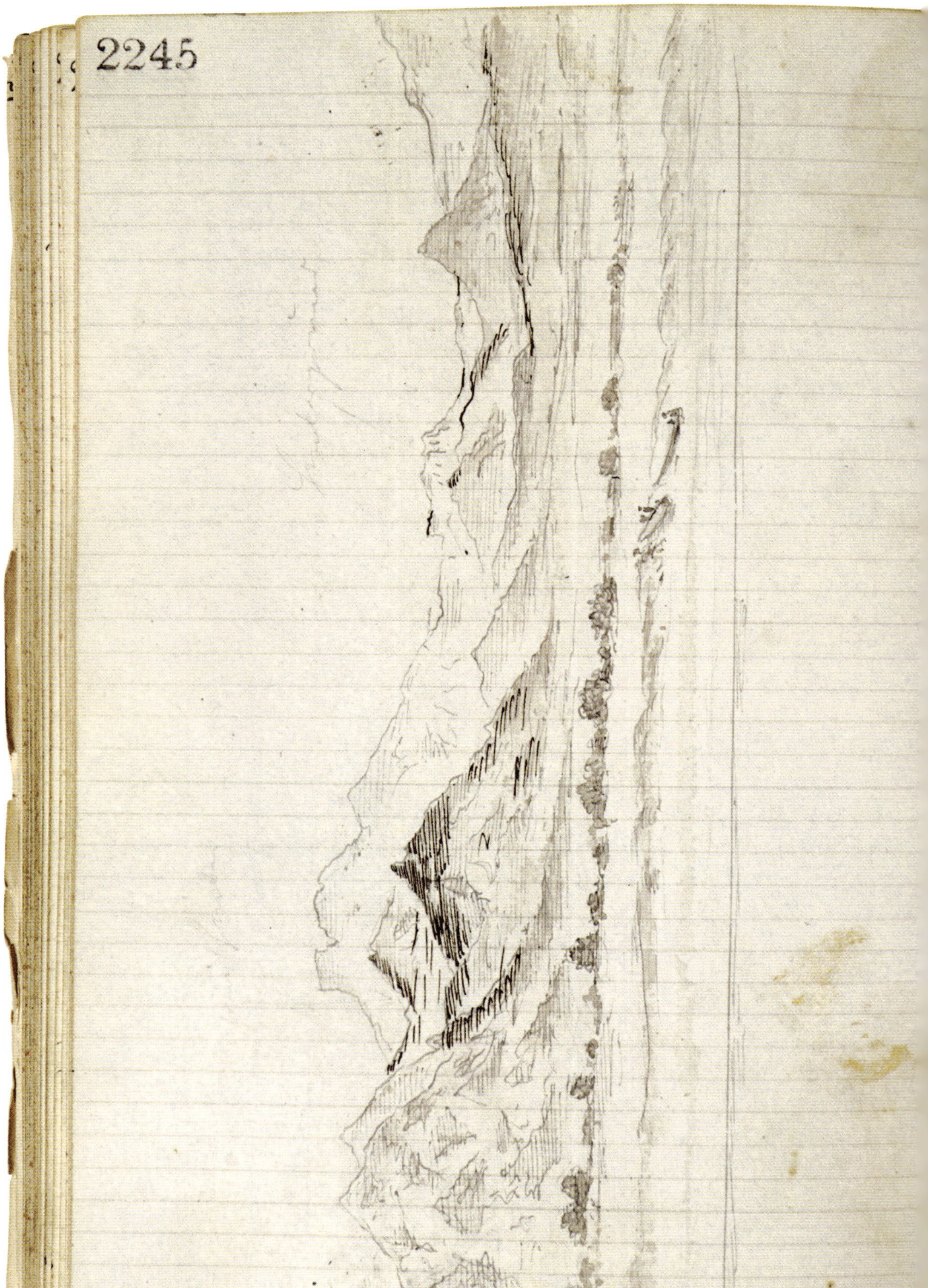
2245

2247

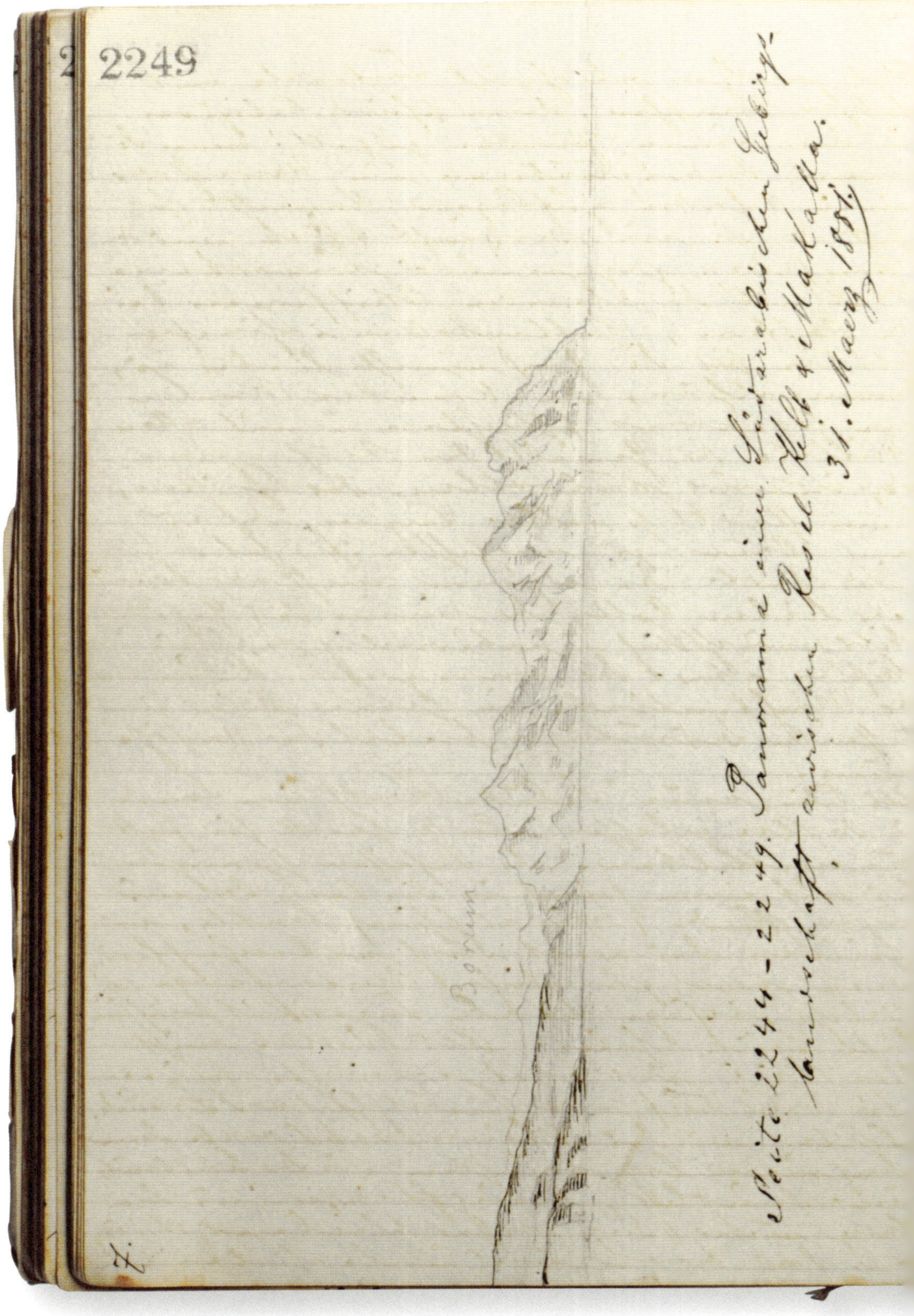
2249
7.
Börum
Seite 2244 - 2249. Panorama einer Südarabischen Gebirgs-
landschaft, zwischen Rasel Kelb & Makalla.
31. Maerz 1881.

Was nützen überhaupt Vorträge mit
solchen Menschen, deren schwache Religion
in dem einen Punkte gipfelt: wenn du
einem Andersgläubigen ein Unrecht oder
einen Schaden zufügen kannst, so
thust du dir ein Verdienst Allah und den
Propheten gegenüber [wie wir] erworben.
Statt all' die theuren Missionen bei
den Kaffern, Hottentotten u. Chinesen
sollten sich die Herrn Missionare lieber zu-
sammenthun, den kranken Mann
am Bosporus schlachten u. mit allen
Mitteln der Gewalt den Muhamme-
danismus ausrotten. Das Christen-
thum strotzt so voller Lügen, daß es
auch kaum zu umfassen ist, hat doch
jede Partei den eigenen mehr oder weni-
ger lieben Gott, der einem das Gesammt-
bild märchenhaft u. erbärmlich erscheinen
läßt – aber der eine große Gedanke
der Nächstenliebe, in seiner alles um-
fassenden Bedeutung bietet die Versöh-
nung.
Als die „neue" Feluke (Boot) vom Lande
in die See gelassen wurde sank sie fast
unmittelbar. Die Insassen sprangen
bis auf einen sofort in die See, der mit
einem Kinde (wie es schien) kauerte, und
hielten das Boot durch unbeschreibliches Schwim-
men vom gänzlichen Untergange ab.
Welche riesige, athletische [illegible] müssen
diese Leute haben. Als Omar mit
dem Rest des Boots versunken schien
rief er auf – [illegible] Freude.
Später aus dem kleinen Boote nach dem
morgigen Tag tauchten die Bootsleute
unermüdlich, [illegible]
lang unter dem Wasser bleibend, [illegible]
[illegible] das Boot und unterhielten sich

Otto Mantey, pen-and-ink drawings of red granite formation in the Haghier Mountains.
Expedition Riebeck / Halle a.d. Saale (Allemagne). / Tagebuch 1881. / Dr Mantey, unpublished MS, pp. 2273–4
The Arcadian Library, 16521

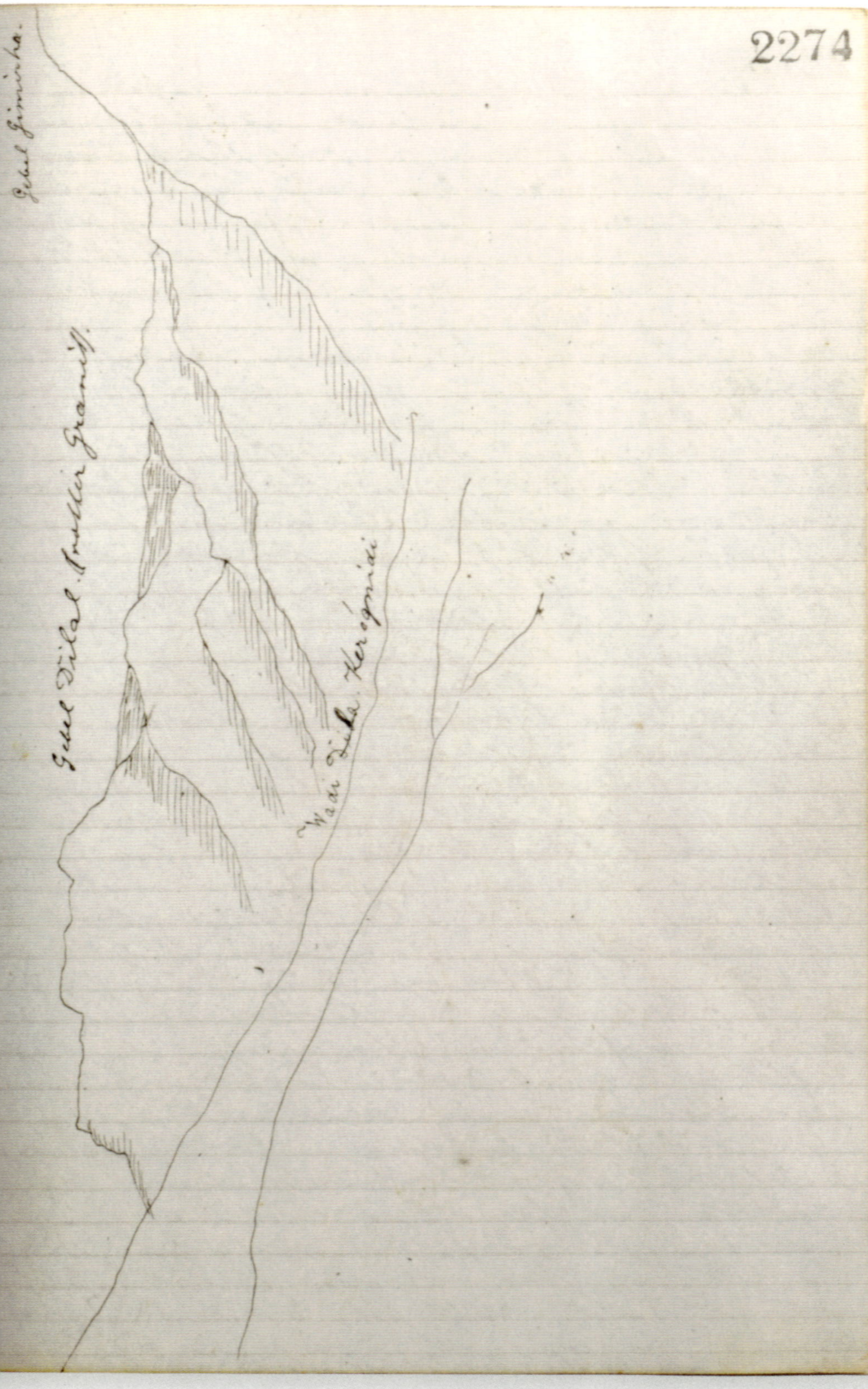

Otto Mantey, watercolour drawing of green vein in red granite formation on the Keregingiti River.
Expedition Riebeck / Halle a.d. Saale (Allemagne). / Tagebuch 1881. / Dr Mantey, unpublished MS, pp. 2281–2
The Arcadian Library, 16521

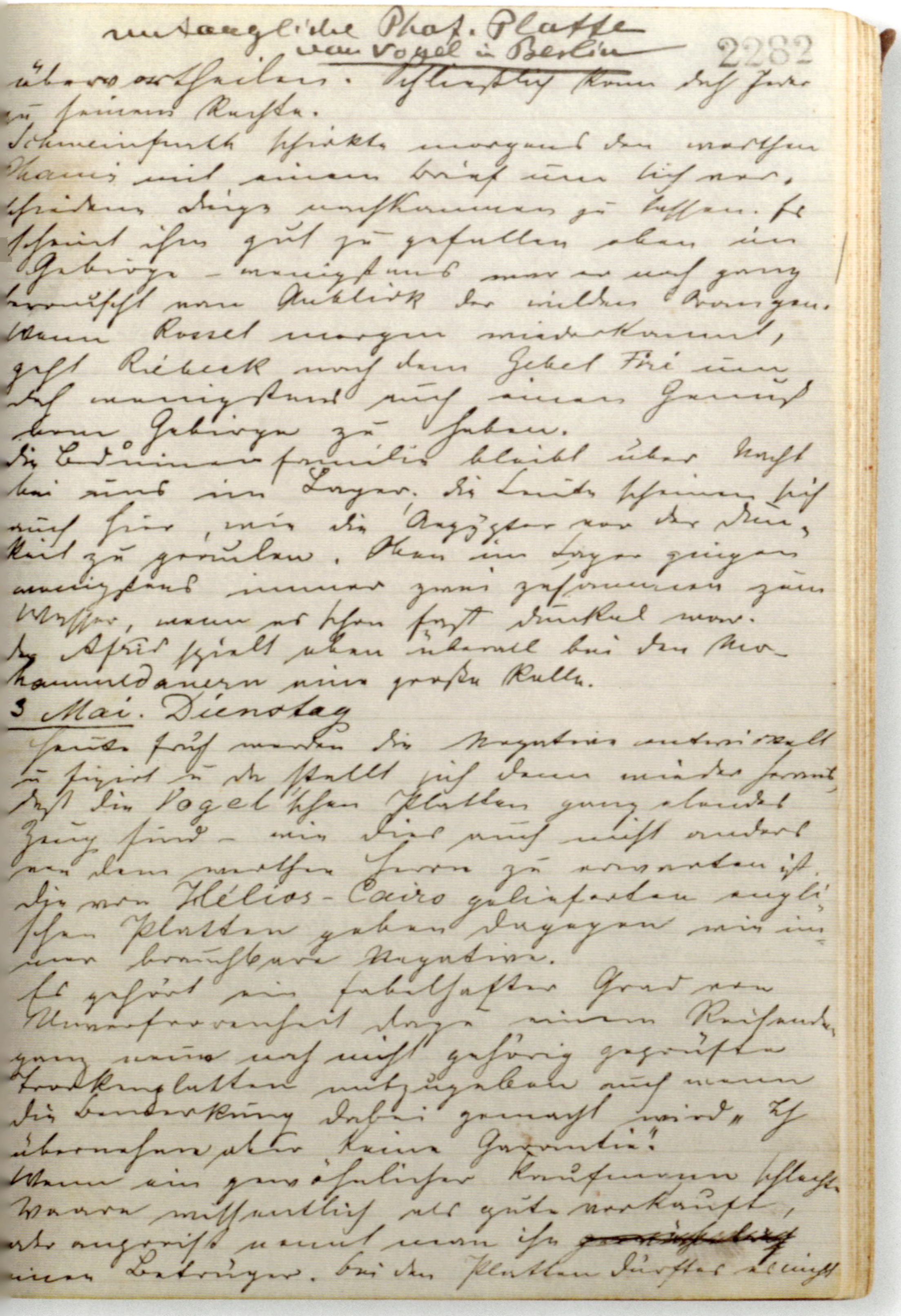
untaugliche Phot. Platte von Vogel in Berlin

2282

über[illegible]. Schließlich kam doch Jeder zu seinem Rechte.
Schweinfurth schickte morgens den weißen Khamis mit einem Brief [illegible], [illegible] dings nachkommen zu lassen. Es scheint ihm gut zu gefallen oben im Gebirge – wenigstens war er noch ganz berauscht vom Anblick der wilden Bergen.
Wenn Rossel morgen wiederkommt, geht Riebeck nach dem Gebel Fîi um doch wenigstens auch einen Genuß vom Gebirge zu haben.
Die Beduinenfamilie bleibt über Nacht bei uns im Lager. Die Leute scheinen sich auch hier, wie die Ägypter vor der Dunkelheit zu grauen. Wenn im Lager gingen wenigstens immer zwei zusammen zum Wasser, wenn es schon fast dunkel war.
Der Afrît spielt eben überall bei den Mohammedanern eine große Rolle.

3 Mai. Dienstag

Heute früh wurden die Negative entwickelt u. fixiert u. da stellt sich denn wieder heraus, daß die Vogel'schen Platten ganz elendes Zeug sind – wie dies auch nicht anders von dem [illegible] Herrn zu erwarten ist. Die von Hélios-Cairo gelieferten englischen Platten geben dagegen wie immer brauchbare Negative.
Es gehört ein fabelhafter Grad von Unverfrorenheit dazu einem Reisenden ganz neue noch nicht gehörig geprüfte Trockenplatten mitzugeben auch wenn die Bemerkung dabei gemacht wird „Ich übernehme aber keine Garantie!"
Wenn ein gewöhnlicher Kaufmann schlechte Waare wissentlich als gute verkauft, oder [illegible] nennt man ihn ~~[illegible]~~ einen Betrüger. Bei den Platten dürfte es nicht

Otto Mantey, pen-and-ink drawings drawing of the mountains east of Hadibo. *Expedition Riebeck / Halle a.d. Saale (Allemagne). / Tagebuch 1881.* / Dr Mantey, unpublished MS, pp. 2287–8
The Arcadian Library, 16521

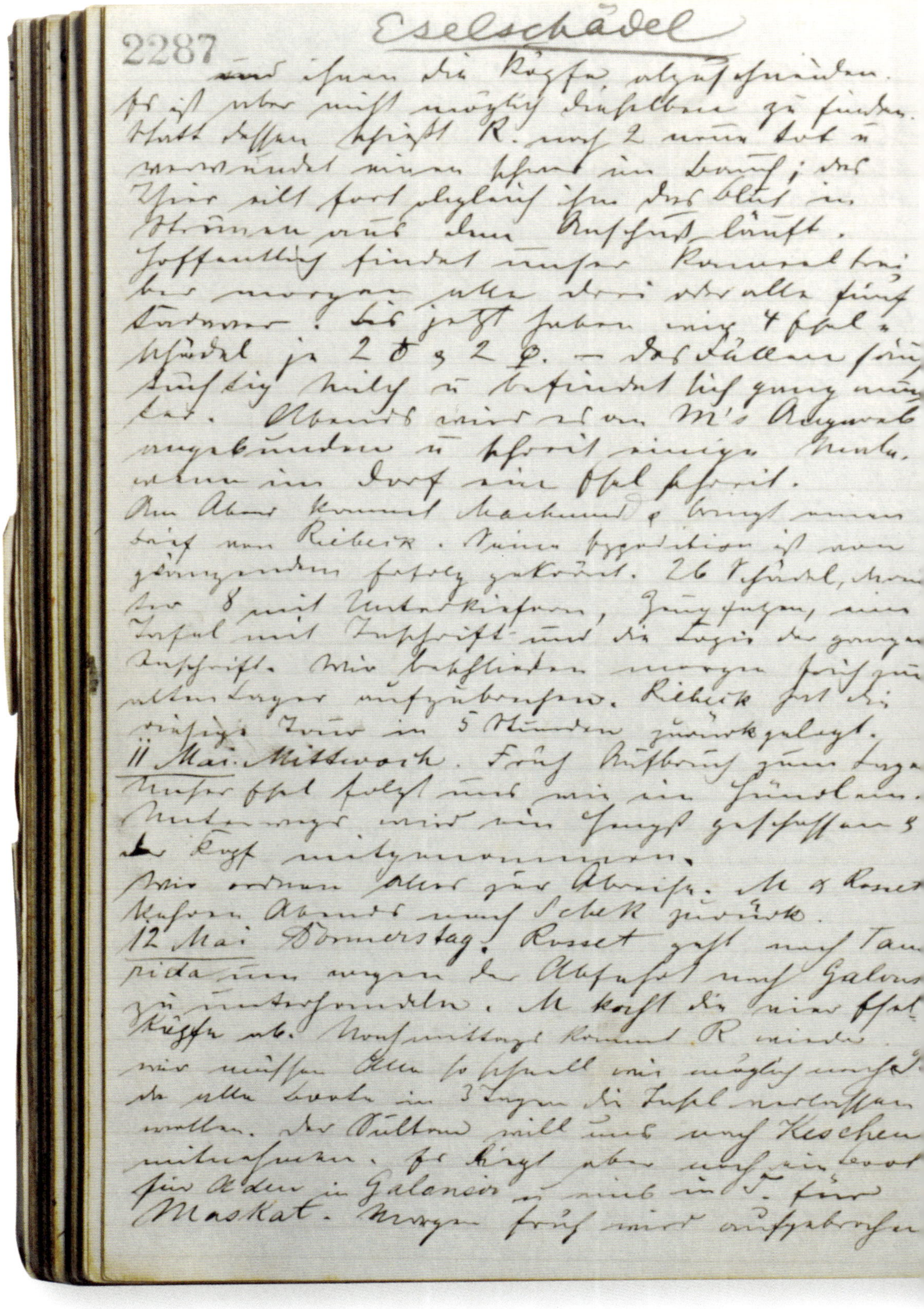

Korallenkalk u. Sedimentgestein
bewachsen
Korallenkalk u. Sedimentgestein.
Gebirge östlich von
Tamarida
13. V. 81.
2288

Georg Schweinfurth, pen-and-ink drawing of cornetfish caught near Mukalla. *Expedition Riebeck / Halle a.d. Saale (Allemagne). / Tagebuch 1881. / Dr Mantey*, unpublished MS, pp. 2295–6
The Arcadian Library, 16521

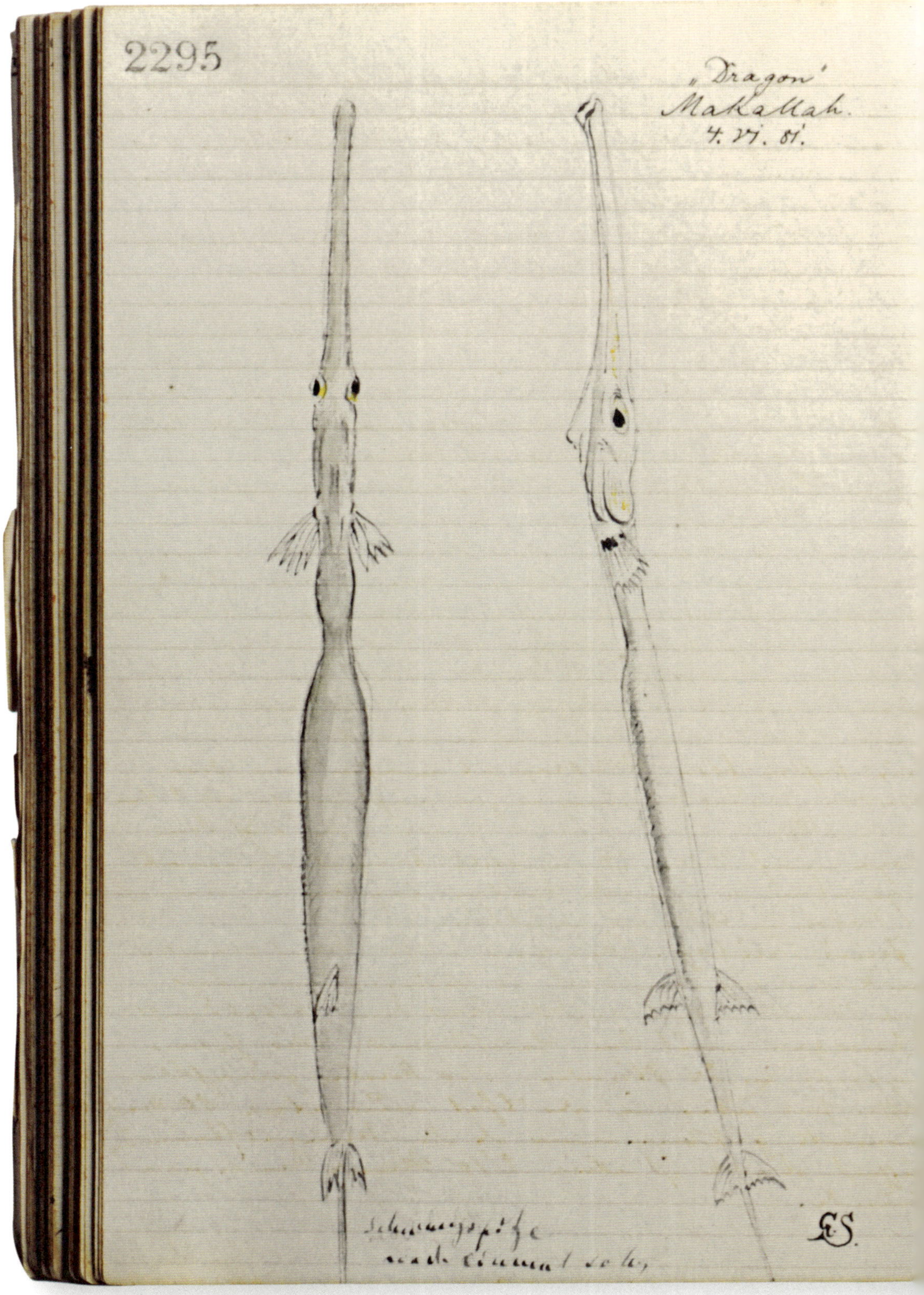

der Horizont wird Nachmittags so trübe, daß wir
icht in Borum ankern können. Wir treiben
unter einigen Segeln durch den Strom S.W. u. sind
am andern Morgen in der Höhe des Tafelberges zwi-
schen Hami & Scheger.

Juni Mittwoch. In Makallal werden neue Schafe
kauft, dann geht es wieder nach Borum.
Nachmittags Schießübungen mit einer kleinen
Kanone (40 Schuß) u. mit Revolvern. Die Kanone
schlägt ein mal ganz um u. verwundet einen
Mann indem sie ihm an der Palmarseite des
rechten Vorderarmes ein handtellergroßes Stück Haut
bis zur Fascia antibrachii losreißt.

Juni Donnerstag. Früh Schießübungen mit den
großen Kanonen. Leider dreht sich das Boot, so
daß nach der schwimmenden Scheibe nur einige
Schüsse gefeuert werden können. Alle anderen Schüsse,
Granaten & Shrapnells werden nach dem Cap [illegible] von
Borum abgegeben. Einige Granaten treffen
die Felsen u. springen dort. – Die großen Kanonen
werden heruntergelassen u. Alles zur Abfahrt für
heute Abend nach Aden vorbereitet.
Wir fahren schon Mittags ab machen aber der
starken Gegenströmung wegen nicht mehr als 4 Knoten.
Gegen 11 Uhr wird ein Handelsdampfer, der auf dem
Wege nach Aden ist mittels Signal herangerufen
angerufen u. ihm der Auftrag gegeben die Ankunft
der Dragon in Aden zu melden.

Juni Freitag. Morgens 5 Uhr. Wind ziemlich
stark, legt sich dann bald. Es werden für ungefähr
eine Stunde Segel gesetzt. Nach dem Gottesdienst
Übungen mit den Geschützen. Wir haben
ungefähr in 24 Stunden, 130 Knoten gemacht im
Wasser, sind aber in der That nur etwa 80 Meilen
vorwärts gekommen da die Meeresströmung
äußerst stark ist etwa 2 1/2 ~~Stunde~~ Meile pro
Stunde.

Otto Mantey, various calculations made during the expedition and a watercolour drawing of a piece of red granite.
Expedition Riebeck / Halle a.d. Saale (Allemagne). / Tagebuch 1881. / Dr Mantey, unpublished MS, inside back cover
The Arcadian Library, 16521

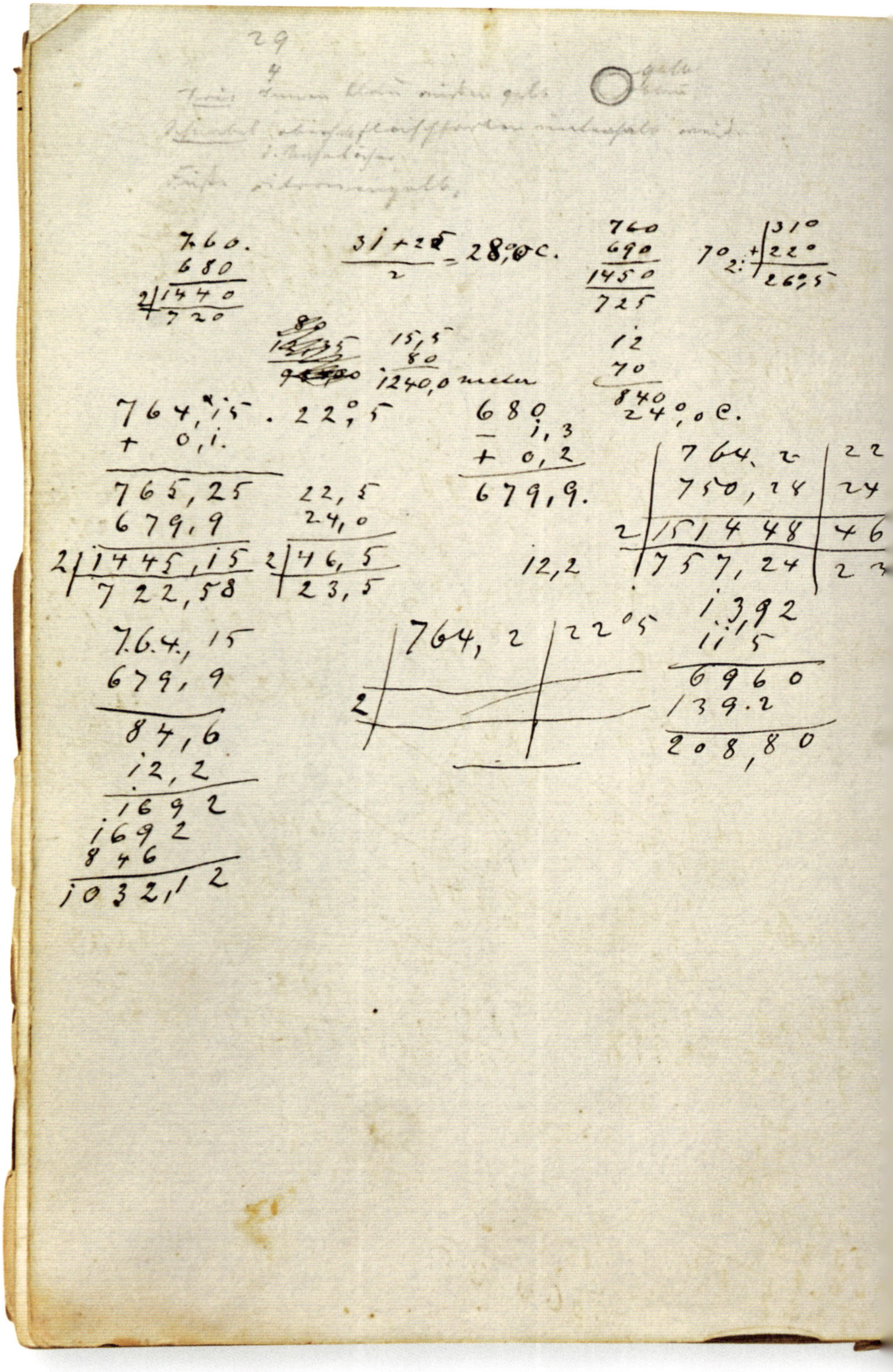

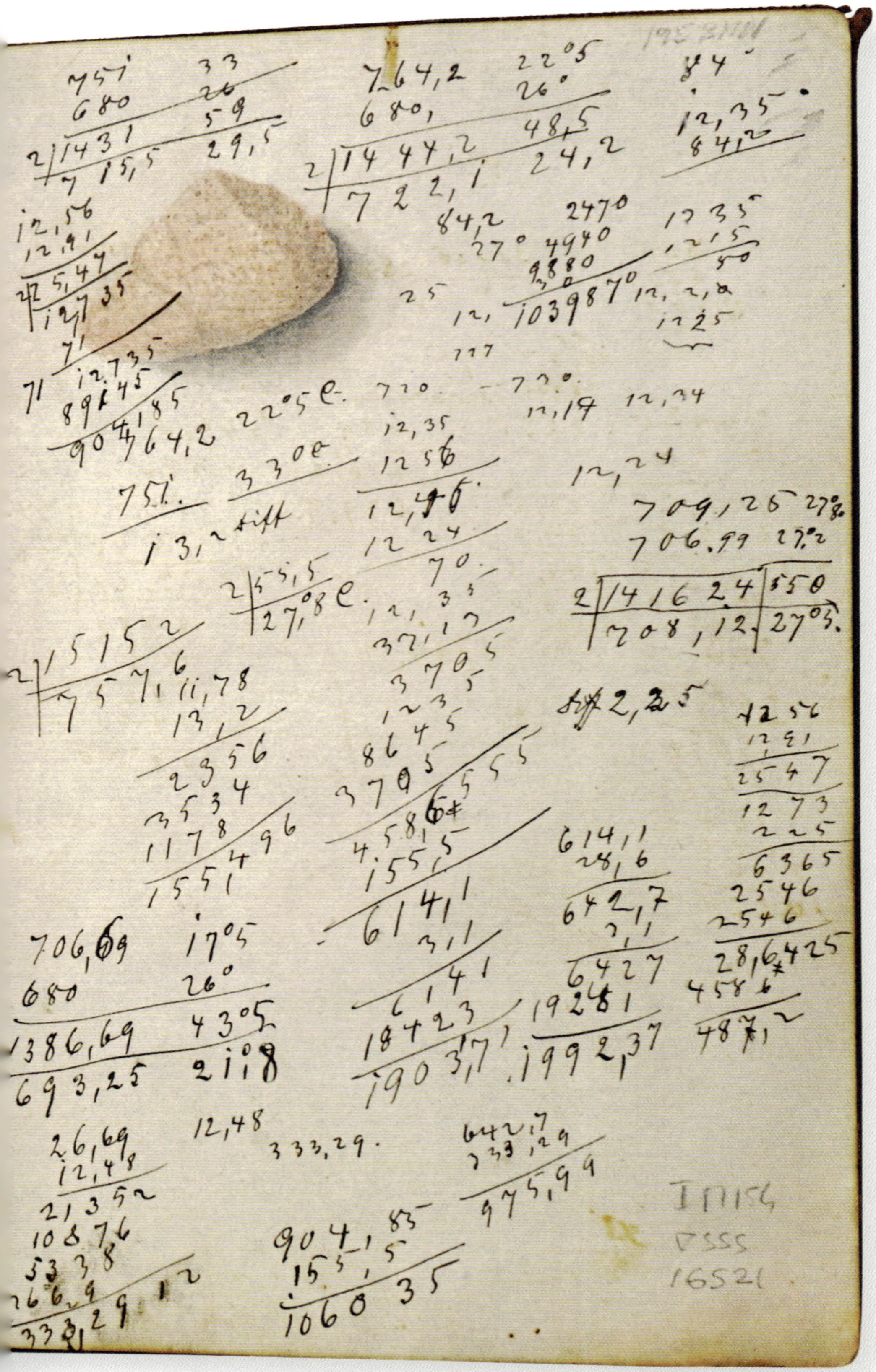

751 33
680 26
2|1431 59
715,5 29,5
764,2 22°5
680, 26°
2|1444,2 48,5
722,1 24,2
764,2 22°5 e.
751. 330 e.
13,2 diff
709,25 27°8
706,99 27°2
2|1416,24|550
708,12|27°5.
706,69 17°5
680 26°
1386,69 43°5
693,25 21°8
904,85
155,5
1060,35

presumably written by Mantey, who substituted for Mook in the expedition. It is illustrated with handsome ink drawings and written in a rough hand.[12] As further proof that this is Mantey's hand, Schweinfurth, in an address of 1883 in Freiburg, loosely quotes from Mantey's report on the fears of the expedition's cook about Soqotran witchcraft, lines which can indeed partially be traced back to the diary.[13]

The newly formed team did not immediately leave for Soqotra. Schweinfurth had for some years been working on a map of the desert between the Nile and the Red Sea, and Riebeck decided to join him on a final survey of the area. For this purpose they organized a small caravan of nine camels, and set off with two servants and a group of ten Bedouin from Tur. The trip was fruitful but, apart from one of the dogs being stung by a wasp, uneventful. The account of this excursion into the Egyptian desert, which is lacking in Mantey's diary, is supplied by Riebeck, who had his diary published in the *Mittheilungen des Vereins für Erdkunde*,[14] including a folding map of Schweinfurth's topographical research of the area.

On 22 February 1881 Riebeck, Schweinfurth, Rosset and Mantey started on the expedition to Soqotra. They travelled first to Suez and from there to Aden. Schweinfurth embarked for Aden on 26 February, while the rest followed on board another steamer two days later. They landed in Aden on 5 March[15] at Steamer Point, and they put up at the Grand Hôtel de l'Univers (where Arthur Rimbaud stayed in the early 1880s). They lost no time before they continued east along the Yemeni coast on board the *Salamati*, beating up against the north-east monsoon. At the coastal town of Shuqrah, about seventy miles north-east of Aden, they cast anchor. If Mantey did not feel there was much to describe in Aden—Rimbaud grumbled that it was 'an awful place without a single blade of grass or a drop of fresh water'[16]—Shuqrah was decidedly more interesting for a naturalist. An excursion on land produced a list of cultivated plants. Mantey also recorded an Umbrella Thorn (p. 2203), which he identified as *Acacia spirocarpa*, and produced a fine sketch of a tree with a flattened crown. However, as this species is limited to eastern Africa, Mantey's acacia is likely to be a specimen of *A. tortilis*, which is typical of the Arabian peninsula, or of *A. campoptila*, a species later described by Schweinfurth for the Yemen.

The expedition's trip along the Yemeni coast was long and, apart from the stormy weather, uneventful. Although there was an occasional jellyfish to dissect (p. 2209), there was little to record and Mantey spent part of his time observing and detailing the workings of their buggalow. One day his attention was caught by an open barrel

12. The rest of the diary is written in this hand, apart from pp. 2291–3, covering the entries for 23–7 May, which are again in the Riebeck/Rosset hand. On their way back to the Yemenite coast Mantey had fallen ill and complained about pains.

13. Cf. Georg Schweinfurth, *Ein Besuch auf Socotra mit der Riebeck'schen Expedition: Vortrag in der zweiten oeffentlichen Sitzung der 56. Versammlung Deutscher Naturforscher und Aerzte zu Freiburg i.B. 1883*, Freiburg i.B. 1884, p. 15 and *Tagebuch*, p. 2259; it is interesting that Schweinfurth apparently had this diary, or a transcript of it, at his disposal when he was preparing his address in 1883.

14. Emil Riebeck, 'Tagebuch einer Reise von Kairo nach den Gebirgen am Rothen Meer', *Mittheilungen des Vereins für Erdkunde zu Halle a.S.*, 1881, pp. 40–50.

15. Although Mantey records: 'Sonntag am 5 März Ankunft in Aden' (*Tagebuch*, p. 2200), 5 March 1881 fell on a Saturday, not on a Sunday.

16. 'Aden est un roc affreux, sans un seul brin d'herbe ni une goutte d'eau bonne.' To his family, 25 August 1880, *Les lettres manuscrites de Rimbaud*, ed. Claude Jeancolas, Paris 1997, p. 411.

which served as a bread oven, while on another he added in his diary an unattractive outline of the vessel. On 26 March they reached Balhaf, one of the four sultanates that made up the Wahidi Sultanate before the 1870s. Rosset and Mantey had time to go on land before they resumed their routine, sailing east. The mountains behind the coastline now caught Mantey's interest, and he included a fine drawing of Maidaha, a small settlement near Ras al-Kelb. In a further series of sketches he traces the sky-line between Ras al-Kelb and Mukalla, which covers six whole pages in the diary (pp. 2244–9). After Ras Sharma they steered south-east and finally crossed the Gulf of Aden. They reached Soqotra on 9 April, landing near Gollonsir before they headed for Tamarida (Hadibo) to pay their respects to the sultan. They pitched their tents on the slopes of Haghier Bitmolech, about 5 miles south of the capital. From there they explored the island for the following six weeks. On 14 May the expedition left Tamarida for Gollonsir, where they hoped to catch the British despatch boat. As there was no sign of any British vessel, and because they wanted to leave the island before the arrival of the south-west monsoon, they had no choice but to embark on an undecked Arab dhow transporting goods to Mukalla.

Although the British and German expeditions to Soqotra were exceptionally short compared to the expeditions described in the first chapters of this book, they were highly successful in exploring the natural history of the island. Balfour writes that, although '[t]he members of the German expedition devoted themselves largely to the study of the people and languages', they still collected many specimens of plants and animals.[17] Although the zoological specimens were sent to Germany for detailed study, Schweinfurth forwarded all his botanical finds to Glasgow. When Balfour heard that a second scientific expedition to Soqotra was on its way, he hastened to send a catalogue of their 'plant-gatherings' after them, so that they could give particular attention to the gaps in their collection. This letter reached Schweinfurth in time and 'with a generosity which is pleasing as it is rare, he subsequently sent his collections to me in England, in order that the whole flora might be worked out in one.' On the value of Schweinfurth's collection Balfour writes: 'In the excellence of his specimens and their completeness, and in the way which they so frequently supplemented, in flower and fruit characters, deficiencies in ours,—therein lay the value of Schweinfurth's plants, and I cannot appraise it too highly.'[18]

An early example in Balfour's *Botany of Socotra* illustrates the importance of Schweinfurth's collection. One of the most remarkable discoveries on the island, according to Balfour, was an endemic hard-wood shrub he found south of Tamarida on the Haghier Hills. On the spot he felt unsure about its affinity, but when he examined the flowers back in Scotland, he was surprised to conclude that it was a new genus in the family of the Menisperms (crescent-shaped seeds). Later in 1881, however, he received Schweinfurth's specimens (carrying 'a quantity of fruit') which the German had recognized as a species belonging to the menisperm genus *Cocculus*, and he realized his mistake. The plant is known today as *C. balfourii* Schweinf. ex Balf.f.

But if this menisperm from the Haghier Hills did not constitute a new genus, many other of Balfour's (and Schweinfurth's) Soqotran finds did. In his *Botany of Socotra* Balfour lists just over 600 entities (not counting taxa below the rank

17. Balfour, *Botany of Socotra*, p. xv.
18. Balfour, *Botany of Socotra*, p. xvi.

of variety), of which about 30 per cent were described as new species. Of these twenty belonged to new genera. What is staggering about these numbers is that the Soqotran flora is estimated to contain a total of about 815 species,[19] and that most of Balfour's new species are considered valid taxa today. These are percentages which are much higher than the ones we have seen for the other scientific expeditions. The explanation of Balfour's success and accuracy in describing new species must no doubt be sought in the fact that Soqotra is an island with a distinct endemic flora.[20] 33 per cent of all species are endemic, so it is hardly surprising that nearly all of Balfour's new species should also be endemic to the island.

Amongst all this wealth of new species (and genera), there was one plant which, Balfour thought, 'in interest surpasses most of the others in the flora'.[21] A common small tree, found by both him and Schweinfurth on the higher parts of the Haghier range, looked conspicuously like a pomegranate. Upon a careful study of its fruit structure, Balfour concluded that its characteristics so markedly '[lead] us back to an earlier stage in the evolution of the type' that it must be 'considered the type of the primitive stock' of our cultivated pomegranate. It was accordingly baptized as *Punica protopunica*.

Another plant singled out by Balfour as 'conspicuously noteworthy' is the famed Dragon's Blood Tree. The Soqotran gum resin of this name is mentioned by Dioscorides in his *Materia medica* as κιννάβαρι, and the first-century *Periplus Maris Erythraei* mentions the island of Soqotra and its produce of dragon's blood:

[...] *it is called Dioscorida, and is very large but desert and marshy, having rivers in it and crocodiles and many snakes and great lizards, of which the flesh is eaten and the fat melted and used instead of olive oil. The island yields no fruit, neither vine nor grain. The inhabitants are few and they live on the coast toward the north, which from this side faces the continent. They are foreigners, a mixture of Arabs and Indians and Greeks, who have emigrated to carry on trade there. The island produces the true sea-tortoise, and the land-tortoise, and the white tortoise which is very numerous and preferred for its large shells; and the mountain-tortoise, which is largest of all and has the thickest shell; of which the worthless specimens cannot be cut apart on the under side, because they are even too hard; but those of value are cut apart and the shells made whole into caskets and small plates and cake-dishes and that sort of ware. There is also produced in this island cinnabar, that called Indian, which is collected in drops from the trees.*[22]

The resin exuding from the stem of this tree had, in fact, been used from antiquity as a medicine for the treatment of eye and skin diseases, for stomach- and headaches, as a coagulant, and for a whole series of other health problems. As a dye it was also used to varnish pottery. The name derived from the legend of a tree which sprung up at the spot where the blood spilled by an elephant and a dragon engaged in fight

19. Cf. 'The Socotra Pages of the Royal Botanic Garden of Edinburgh', http://rbg-web2.rbge.org.uk/soqotra/history/page06.html#Flora (last visited 12 July 2007).

20. Cf. Frank White, *The Vegetation of Africa, a Descriptive Memoir to Accompany the UNESCO/AETFAT/UNSO Vegetation Map of Africa*, UNESCO 1983.

21. Balfour, *Botany of Socotra*, p. xxxvii.

22. W.H. Schoff, *The Periplus of the Erythraean Sea: Travel and Trade in the Indian Ocean by a Merchant of the First Century*, London, Bombay and Calcutta 1912, §30.

mingled and congealed in the red beads of cinnabaris.

Although other dragon trees had long since been described, notably the *Dracaena draco* L. of Teneriffe, Dioscorides' Soqotran tree had not. Wellsted had described the tree in an article of 1835, but he erroneously identified it as the West Indian dragon's blood tree *Pterocarpus draco* L. Balfour therefore published it in 1882 as a new species, *D. cinnabari*, adopting the old Greek name. Schweinfurth was clearly not as attracted by the Dragon's Blood Tree as Balfour. In his paper on Soqotra, read to the meeting of naturalists and physicians in Freiburg in 1883, he did not waste many words on the Soqotran dracaena, which he thought only little different from the dragon's blood tree of the Canary Islands.[23] Balfour, however, saw the two species as 'quite distinct one from the other' and, in the species description, repeated that he thought it 'one of the most interesting plants of the island'.[24] He even inserted a picturesque plate of the tree, complete with two Soqotrans, based on a drawing by Lieutenant Cockburn, among the analytical botanical plates of his *Botany of Socotra*.

The zoological results of the two expeditions were as spectacular as the botanical achievements. Balfour reported that the island teemed with insect life, that birds, lizards and land molluscs were plentiful, and that the rivers were stocked with fish and crabs.[25] Although not as exhaustive as the botanical collections, the zoological finds still led to a number of important discoveries. As with the flora, the fauna represented a high percentage of endemic species.

Balfour reported finding three species of snakes on the island, a new genus, a new *Zamenis* species, and a viper known from the Arab mainland.[26] In 1881 A. Günther described the new genus as *Ditypophis*, and the species he described as *Z. socotrae* has recently been raised to the type of another new endemic genus: *Hemerophis* (2001). The status of Balfour's viper, however, remains controversial. No vipers have ever been found on the island and there are no reports of bites among the local population. Balfour's viper might well refer to a specimen found on the Yemeni coast. The members of the Soqotra expeditions also found that gecko and lizard species were common, and their collections yielded six new entities. A particularly large gecko was found by Riebeck and was described in 1882 by Peters as *Haemodracon riebeckii*. It is an endemic genus.

Balfour's reference to fresh-water fish is puzzling. Schweinfurth wrote that there are no true freshwater fish in the island,[27] and subsequent research has confirmed this. The absence of fish might be due to the massive presence of endemic species of crab in the rivers.

Except for the Lepidoptera collected during the British expedition, and described in an article by Arthur Butler,[28] Balfour did not give any information about the island's insect life. It is not clear whether any insects were collected and where they ended up. The ones collected by the Riebeck expedition, however, were handed

23. Schweinfurth, *Ein Besuch auf Socotra*, p. 18.
24. Balfour, *Botany of Socotra*, p. 293.
25. Balfour, *Botany of Socotra*, p. xxx.
26. In his list he also includes a new *Amphisbaenia* (worm lizard) genus.
27. Schweinfurth, *Ein Besuch auf Socotra*, p. 10.
28. Arthur G. Butler, 'On the Lepidoptera Collected in Socotra by Prof. I. B. Balfour', in *Proceedings of the Zoological Society of London*, 1881, pp. 175–80.

Cockburn del.

Dracæna Cinnabari, *Balf. fil.*

J. N. Fitch imp.

over to Otto Taschenberg, who described a handful of new species from them.[29] Taschenberg also published Franz Hilgerdorf's description of the fresh-water crab *Telphusa socotrensis* (=*Socotrapotamon socotrensis*), based on the approximately 30 animals Riebeck had collected near Hadibo.[30] In all Balfour collected 24 butterflies. Butler found that they belonged to thirteen different species and that half (7) of these were new to science. He dedicated to Balfour a big *Charaxes* species with a wing span of well over three inches: *C. balfouri*. Seeing that almost two hundred species of butterflies are now known for Soqotra (and about a third endemic), Balfour's results here are disappointing, even if one takes into account that collecting was restricted to only a few weeks in spring.

On Soqotra Balfour collected 126 specimens of birds, mostly skins, referable to 36 species, while the German expedition added another seven species to this. As might be expected, many of these were widely disseminated outside the island, but Sclater and Hartlaub described seven, mainly endemic, new species.[31] Balfour captured seven specimens of a sunbird which was common in the interior of the island and had a nest with a lateral opening on a small pendent branch. The species was published in the *Proceedings of the Zoological Society of London* and given the name *Cinnyris balfouri* (=*Chalcomitra balfouri*). A fine plate illustrates this endemic bird with its nest. The other bird at the bottom, the Soqotra Warbler, is also an endemic species.

Soqotra is not rich in mammal species. Schweinfurth drew attention to the fact that, although the island's natural conditions ('Daseinsbedingungen') would also seem to favour the presence of such mammals as antelopes, hares, foxes, hyenas, leopards, and Rock Dassies, none of these can be found there.[32] As Soqotra is only very thinly populated by humans, he thought it unlikely that these animals were depleted by hunting. Balfour, too, in the introductory chapter to *Botany of Socotra* underlined the island's 'striking [...] paucity of indigenous mammals'.[33] Oddly enough, he did not mention the presence of the wild ass, the ancestor of the domestic donkey. Although he tried to bring two live civet cats home, he did not seem to have contemplated transporting a live wild ass to Europe. Mantey's diary, on the other hand, relates that Riebeck shot a number of individual specimens for study, and a watercolour of the animal was included in the diary's last pages.[34] The presence of *Equus africanus* on Soqotra was indeed deemed of some interest. Once found across northern Africa, by the 1880s the wild species was limited to Somalia, Ethiopia and eastern Sudan.[35] A Nubian and a Somalian subspecies can be distinguished, the latter having horizontal stripes across the legs, not unlike those of the zebra. The drawing in Mantey's diary shows a typical specimen with unstriped legs which must be considered a feral donkey.

Intrigued by the presence of the wild ass on Soqotra, Schweinfurth dedicated a few paragraphs to the animal. While he was convinced that it was a genuine

Charles Cockburn, Dragon's Blood Tree with two Soqotrans.
Isaac Bayley Balfour, *Botany of Socotra*, Edinburgh and London 1888, plate XCVI

OVERLEAF LEFT
Edwin Wilson, lithograph with Soqotran butterflies with *Charaxes balfouri* in the centre.
Arthur G. Butler, 'On the Lepidoptera collected in Socotra by Prof. I. B. Balfour', *Proceedings of the Zoological Society of London* (1881), plate XVIII
Collection J. M. I. Klaver

OVERLEAF RIGHT
J. Smit, lithograph of the Soqotran Sunbird with nest (top) and the Soqotran Warbler (bottom).
Philip Lutley Sclater and Gustav Hartlaub, 'On the Birds collected in Socotra by Prof. I. B. Balfour', in *Proceedings of the Zoological Society of London* (1881), plate XV
Collection J. M. I. Klaver

29. Only his beetle *Pachysmopoda abbreviata* and his grasshopper *Systates angusticollis* are accepted names.
30. Otto Taschenberg, 'Beiträge zur Fauna der Insel Sokotra, vorzüglich nach dem Hernn Dr. Emil Riebeck aus Halle a.S. gesammelten Materiale zusammengestellt', in *Zeitschrift für Naturwissenschaften*, 1883, pp. 171–3.
31. Sclater and Hartlaub's *Lanius uncintus* is today considered a subspecies of *L. meridionalis*.
32. Schweinfurth, *Ein Besuch auf Socotra*, p. 10.
33. Balfour, *Botany of Socotra*, p. xxx.
34. *Tagebuch*, pp. 2285–7, 2378.
35. Today the wild ass is classified as 'critically endangered' on the *IUNC Red List*.

Edwin Wilson del. et lith. Mintern Bros. imp.

SOCOTRAN BUTTERFLIES.

J. Smit lith. Hanhart imp.

I. CISTICOLA INCANA. 2. CINNYRIS BALFOURI *with nest.*

P.Z.S.1881.Pl.XVII.

J. Smit lith.

Hanhart imp.

RHYNCHOSTRUTHUS SOCOTRANUS.

wild species, he could not but conclude that, at some stage, the animal had been introduced to the island by men. If the possibility that at one point in history it had swum across the 220 kilometres that divide Soqotra from Somalia needed to be excluded a priori, the theory of the existence of a former land bridge would have resulted in the presence of other African mammals as well. He therefore concluded that the Soqotran wild ass was most likely the result of imported animals which, through time, had reverted back to the wild type.[36]

J. Smit, lithograph of *Rhynchostrutus socotranus*. Philip Lutley Sclater and Gustav Hartlaub, 'On the Birds collected in Socotra by Prof. I. B. Balfour', *Proceedings of the Zoological Society of London*, 1881, plate XVII
Collection J. M. I. Klaver

Despite the botanical, zoological and ethnological success of the expeditions, neither the British nor the Germans were sorry to leave the island. When the despatch boat of the Bombay Marine arrived to pick up the British naturalists, at the end of March 1880, Balfour commented: 'We were not sorry to see her, as our camp was now very sickly—Scott was down with the fever, one coolie had sunstroke, and the other servants were all suffering badly from fever.'[37] The Germans too, when confronted with the possibility of remaining trapped on the island for a year because of the impending monsoon season, opted for an uncomfortable journey back to the mainland on an Arab vessel rather than risking isolation from the rest of the world for months. The voyage towards the Yemeni coast was a difficult one, the winds being unfavourable, and Mantey fell ill.[38] In Mukalla they had counted on help to obtain camels for a caravan to Aden. Sultan Nakib Omar, however, had broken all diplomatic relations with the English (and other European nations) and threatened war. Their water supply nearly exhausted, the vessel was forced to sail east along the coast to Sheher. Here they were invited on board the *Dragon*, an English man-of-war which was blockading Mukalla. Notwithstanding the hospitality of the English, the naturalists had to make the best of a tedious three-week stay on an overloaded vessel. Mantey's diary entries are short and uninspired. Schweinfurth spent some of his time embellishing the diary with a drawing of a cornetfish caught off the coast of Mukalla (p. 2295). They reached Aden on 12 June 1881. Schweinfurth returned to Cairo, while Riebeck, Rosset and Mantey boarded a steamer for Bombay.

Although both the British and German visits to Soqotra were scientific expeditions, the imperial implications of the island's strategic geographical position were never forgotten. When Captain Haines and Lieutenant Wellsted surveyed Soqotra during the mid-1830s, this was first and foremost to ascertain whether it was suitable as a coaling station on the trade route to India. At the time the British Government tried to buy the island, but failed to do so. In 1839 it seized Aden instead, and lost interest in Soqotra till the mid-1870s when, as Balfour puts it, 'a prospect of its being occupied by another European nation caused the British Government to turn attention to Soqotra, with the result that in that year [1876] a treaty was concluded with the Sultan, by which he binds himself, and his heirs [...] never to cede Socotra to a foreign power, or to allow a settlement on it without consent of the British Government.'[39] Balfour wrote this in 1888, but in 1883 Schweinfurth had already wryly commented in his address in Freiburg that the

36. 'Ueber das wilde Bürgerrecht der Esel auf Socotra kann heutzutage nicht der geringste Zweifel obwalten.' Schweinfurth, *Ein Besuch auf Socotra*, p. 10.
37. Balfour, *Botany of Socotra*, pp. xii–xiii.
38. The *Tagebuch* is continued in a different hand.
39. Balfour, *Botany of Socotra*, p. xxvi.

English were thinking of Germany when they spoke of a 'foreign power'.[40] That people back home in Germany were still interested in the potential colonial value of Soqotra is indicated by the conclusion of Schweinfurth's address:

Otto Mantey, hand-coloured drawing of feral donkey on Soqotra.
Expedition Riebeck / Halle a.d. Saale (Allemagne). / Tagebuch 1881. / Dr Mantey, unpublished MS, p. 2378
The Arcadian Library, 16521

From what I have said it is clear that making Socotra into a colony is meaningless, unless all physicians decided to reintroduce Aloe as an important ingredient in their prescriptions. Then Socotra would earn millions and its population would earn a living again as Aloe collectors, as they did in the times of Alexander.[41]

Despite the humour of this passage, the fact that these lines conclude a paper communicating the results of a scientific expedition is indicative. On the other hand, from this perspective, the British–German collaboration in working out the scientific results of their respective Soqotra expeditions becomes even more remarkable.

40. Schweinfurth, *Ein Besuch auf Socotra*, p. 18.

41. 'Aus dem Gesagten ergibt sich, dass von einer Colonie auf Socotra nicht die Rede sein kann, es sei denn alle Aerzte der Welt kämen überein, der Aloë wieder den alten Ehrenplatz in ihren Recepten einzuräumen. Dann könnte Socotra freilich Millionen daran verdienen und eine ganze Bevölkerung von Aloësammlern und Aloëpflanzern fände wieder, wie in den Tagen des grossen Alexander, auf der Insel ihr Brod.' Schweinfurth, *Ein Besuch auf Socotra*, p. 19.

CONCLUSION

HIGH HOPES, DISILLUSIONMENT AND FINAL ACHIEVEMENTS

Alire Raffeneau-Delile, plate of *Nymphaea lotus* and *Nymphaea caerulea* (detail).
Étienne Geoffroy Saint-Hilaire, Jules-César Savigny *et al.*, *Description de l'Égypte*, Paris 1809–12, *Histoire naturelle*, vol 2 bis, Botanique, plate 60
British Library, 1899.k.1

The naturalists and scholars who set out for the Near and Middle East were aware they were embarking on important scientific missions and counted on being congratulated on their success when they came home with their discoveries. Carsten Niebuhr's son reminded his readers that Von Haven 'always talked of a comfortable future after his return',[1] and Forsskål agreed to become part of the Danish expedition on condition that he was given the title of professor.[2] Geoffroy half jokingly wrote to Cuvier that, after his discoveries in Egypt, he 'pretend[ed] to no less than the throne of anatomy'.[3] Cuvier, who had already achieved fame, on the other hand, gracefully declined the invitation to go to Egypt with Bonaparte. Ehrenberg 'expect[ed] *Eclat* from our hard work',[4] and was therefore unwilling to return to Germany before his work was thoroughly done. He feared the derision of the world of learning and that scientists abroad would say that 'those horrible Germans cannot even bring a voyage to an end'.[5] Academic recognition would have justified an experience which was not without danger. Yet many of the young men who left for Egypt, the Red Sea and the Yemen never returned. Forsskål, Von Haven, Baurenfeind, Kramer, Coquebert and Hemprich all succumbed to disease. And those who survived—Niebuhr, Geoffroy, Savigny, Delile and Ehrenberg—came home older and sadder men.

The various fortunes of the collections, the fruit of so much suffering and hardship, were by no means smooth either. Much of Forsskål's collection was lost during transport back to Europe, while a considerable part of the specimens that reached Copenhagen was so severely damaged that it had become worthless for study. Moreover, by the time Niebuhr returned, Michaelis' Augustus of the North, King Frederick V, had died, and his son, the seventeen-year-old Christian VII, had little interest in the results of the expedition his father had sponsored. Hansen also tells how the chests which reached Denmark were stored unopened and were not inspected until 1882, when Morten Thrane Brünnich replaced Ascarius as director of the Natural History Museum in Copenhagen.[6] Unfortunately Forsskål's notes were also edited in a very unsatisfactory way. The anonymous editor and naturalist whom Niebuhr employed for the publication of Forsskål's work used a highly irregular method of compiling the lists and descriptions of species, introducing numerous inconsistencies and making the works impracticable to consult. All this has led to the sombre view of Forsskål's achievements that Hansen ultimately adopts. In his final assessment of the natural history part of the Danish expedition he quotes Carl Christensen, who wrote in an early biography: 'If one takes the view that an historical account of an individual and his works is of interest only in so far as he can

1. Barthold Georg Niebuhr, 'Carsten Niebuhrs Leben', I.894.
2. Hansen, *Arabia Felix*, p. 29.
3. Geoffroy to Cuvier, 16 August 1799, Geoffroy, *Lettres écrites d'Égypte*, p. 129.
4. Geoffroy to Hemprich, 3 January 1824, Stresemann, *Hemprich und Ehrenberg*, p. 108.
5. Hemprich to Lichtenstein, 16 September 1822, Stresemann, *Hemprich und Ehrenberg*, pp. 69–70.
6. Hansen, *Arabia Felix*, pp. 277–8.

be seen to have contributed to the development of science, then in this account of Forsskål there is little of value.'[7]

The French collections fared better. Notwithstanding Geoffroy's fears that the spirits in which he had conserved his specimens might evaporate with the prolonged delay in Alexandria before the naturalists were allowed to return to France, and that ultimately an expedition which set out with the fairest of prospects had been ruined, most of the naturalists and their collections reached France safely. Once he had returned to France, however, Geoffroy showed little interest in the prestigious publication that was to include all the discoveries of the Egyptian expedition. Of the zoological sections that fell to his part he only finished a few chapters before his scientific interests led him in other directions, and he left most of the work to be completed by his son. The other zoologist of the expedition, Savigny, worked industriously for the *Description de l'Égypte* between 1805 and 1814, and supervised many of the plates for the sections on birds and invertebrates. But he was struck down by a neurological disorder which compelled him to live in almost complete darkness and which made it impossible for him to write the texts for the plates. These were written by Victor Audouin, a young naturalist who had never been to Egypt. He was commissioned to describe and identify the species in the illustrations without consulting Savigny. This scientifically unsatisfactory arrangement explains the many imperfections and mistakes that can be found in the texts that accompany Savigny's plates.

The work of the naturalists of the Prussian expedition was frustrated from the very beginning. Complaints about disease and lack of funds recur in their narrative. But the expedition was above all deeply affected by Hemprich's death. Without his friend's personal experience and knowledge, Ehrenberg was unable to work out on his own the wealth of material they had sent back to Europe. Besides, Ehrenberg discovered to his chagrin that the collections had been severely tampered with in Berlin and that many specimens had been sold to other natural history collections in Europe. It is hardly surprising that he lost interest in the project of publishing all their discoveries systematically. Within a few years of his return to Europe he accompanied Alexander von Humboldt on an expedition to Russia, and later in life dedicated himself mainly to the study of micro-organisms. In this respect the aftermath of the Prussian expedition was not unlike that of the French one. The *Symbolae physicae*, the book which was to rival the French *Description de l'Égypte*, was never finished.

But these negative assessments are only part of the story. Despite loss of life and personal disillusionment, the Danish, French and Prussian expeditions were highly successful in mapping out a flora and fauna which had hitherto remained mostly unknown. The *Flora Aegyptiaco-Arabica*, based on Forsskål's notes, contains just over 2,100 species. Of the almost 1,200 species listed for Egypt and the Yemen a fifth proved to be new. And, as the Herbarium Forsskålii in Copenhagen contains 1,750 sheets with approximately 1,200 species and varieties, it is clear that a considerable part of Forsskål's botanical collection reached Europe safely after all and that his collecting had not been in vain. Much the same can be said about Forsskål's collection of animals. In his *Descriptiones animalium* 641 species are listed and the main section

7. Hansen, *Arabia Felix*, p. 282.

of the book provides descriptions of a great number of birds (17), reptiles (13), fish (151), insects (62), worms (60) and corals (26).[8] Much more of Forsskål's collection of animals ultimately reached Denmark than Hansen supposed, and it has proved an important source for numerous type specimens of species.

The monumental *Description de l'Égypte* testifies to the success of the French expedition. Its natural history volumes are the first fully illustrated study of the Egyptian flora and fauna. His weak health notwithstanding Savigny published numerous new genera and species, mainly of birds and invertebrates, as well as the famous Egyptian lotus (*Nymphaea caerula*). Geoffroy described a remarkable number of new bat species and did important work on the fish of the Nile. Delile listed about one thousand species for Egypt, almost doubling the number of species included in Forsskål's *Flora Aegyptiaco-Arabica*.

The staggering quantity of specimens sent home by Ehrenberg and Hemprich is indicative of the ultimate success of the Prussian expedition. Coming after two other major scientific expeditions, during which many new species were described by Forsskål and the French naturalists (a fact about which Ehrenberg repeatedly grumbled during his travels), an impressive number of new genera, species and varieties was described in *Symbolae physicae*. This is all the more remarkable when one considers that Ehrenberg soon lost interest in the project and never completed his work on the Middle East.

Compared to the Danish, French and Prussian expeditions, the explorations of the flora and fauna of Soqotra by Balfour and Schweinfurth in the early 1880s are only minor events. Still, the great number of unique, endemic species they found on the island made their discoveries of scientific interest. They further completed the investigation of the natural history of the region by visiting the area's only major island.

It is clear that the final assessment of the success of the late eighteenth-century and early nineteenth-century expeditions exploring the natural history of the Near and Middle East must be positive. The joint efforts of Forsskål, Geoffroy, Savigny, Delile, Hemprich and Ehrenberg laid the foundation for a modern taxonomical study of the flora and fauna of Egypt, the Red Sea and the Yemen. Taxonomists today still need to return to their descriptions and specimens for species that can be found in a large part of Africa and Asia. Moreover, the number of species described by the explorers themselves are not the only measure of the success of the expeditions. The results published by Niebuhr, Geoffroy, Savigny and Ehrenberg may well have suffered from various shortcomings, but numerous new species were ultimately described from their collections by later naturalists, many of whom never visited the Arab world.

Finally, the great scientific expeditions of the eighteenth and nineteenth centuries are important landmarks in the history of botanical and zoological illustration. The plates in Forsskål's *Icones rerum naturalium*, based on Baurenfeind's drawings, are early examples of scientific precision. Redouté's drawings, which provided exemplars for the plates in the *Description de l'Égypte*, are justly admired for both exactness and artistic skill, while the coloured plates in Ehrenberg's *Symbolae physicae* added a

8. Cf. Torben Wolff, 'De zoologiske resulater af Den Arabiske Rejse', in *Den Arabiske Rejse 1761–1767*, ed. Stig Rasmussen, Copenhagen 1990, p. 233.

Romantic liveliness to their subjects. As for the written histories of the expeditions, whether in diaries, letters, or official publications, they play an important part in the history of European exploration. And, despite the personal suffering and hardship, these splendid writings testify to the emotions the naturalists felt when they studied the unknown flora and fauna of the area—emotions similar to those reported by Charles Darwin in his concluding remarks about his travelling experiences in the early 1830s: 'I do not doubt that every traveller must remember the glowing sense of happiness he experienced, from the simple consciousness of breathing in a foreign clime.'[9] The geologist Charles Lyell was more eloquent still on the sense of gratification one feels when unravelling the plan of nature. Commenting on the results of his own explorations he pointed out that, although complete knowledge was still imperfect, '[m]eanwhile the charm of first discovery is our own, and as we explore this magnificent field of enquiry, the sentiment of a great historian of our times may continually be present to our minds, that "he who calls what has vanished back again into being, enjoys a bliss like that of creating."'[10] The quotation is of special interest here, since the historian Lyell referred to was none other than the author of Carsten Niebuhr's *Life*, his son Barthold Georg Niebuhr.

9. Darwin, Charles, *Narrative of the Surveying Voyages of His Majesty's Ships Adventure and Beagle between the Years 1826 and 1836, Describing their Examination of the Southern Shores of South America, and the Beagle's Circumnavigation of the Globe. Journal and Remarks. 1932–1836.* London 1839, p. 606.

10. Lyell, Charles, *Principles of Geology: Being an Attempt to Explain the Former Changes of the Earth's Surface, by Reference to Causes now in Operation*, 3 vols., London 1830–3, I.74.

BIBLIOGRAPHY

AL-JABARTI, ʿABD AL-RAHMAN, *ʿAbd al-Rahmān al-Jabarti's History of Egypt*, ed. Thomas Philipp and Moshe Perlmann, Stuttgart 1994.

Allgemeine Deutsche Biographie, 56 vols., 1875–1912: 'Ehrenberg, Christian Gottfried', in *Elektronische Version* 5, pp. 701–11.

ANON., 'Dr. Emil Riebeck', *Saale Zeitung*, 144, 24 June 1885.

ANON., 'Obituary: Dr. Emil Riebeck', *Proceedings of the Royal Geographical Society and Monthly Record of Geography*, 7, August 1885, p. 545.

ANON., 'Dr. Emil Riebeck', *Illustrierte Zeitung*, 20 September 1884, p. 281.

ANON., 'Proceedings of Foreign Societies: Dr. Schweinfurth's Letters Read before the Geographical Society of Berlin, 17 July 1881', *Proceedings of the Royal Geographical Society and Monthly Record of Geography*, 3, September 1881, p. 572.

BALFOUR, Isaac Bayley, *Botany of Socotra*, Edinburgh and London 1888.

BAUER, Aaron M., 'The *Symbolae physicae* and the Herpetology of Hemprich's and Ehrenberg's Expedition to Egypt and the Middle East', *Newsletter and Bulletin of the International Society for the History and Bibliography of Herpetology*, 2, 2000, pp. 8–16.

BEAUCOUR, Fernand, Yves LAISSUS and Chantal ORGOGOZO, *The Discovery of Egypt: Artists, Travellers and Scientists*, Paris 1990.

BELON DU MANS, PIERRE, *Les observations de plusieurs singularitez et choses memorables, trouvées en Grèce, Asie, Iudée, Égypte, Arabie, et autres pays estranges*, 3 vols., Paris 1553.
— *Portraits d'oyseaux, animaux, serpens, herbes, arbres, hommes et femmes, d'Arabie et Égypte*, Paris 1557.

BERNARD, J. F., *Talleyrand: A Biography*, London 1973.

BERNOYER, François, *Avec Bonaparte en Égypte et en Syrie 1798–1800*, Paris 1981.

BLUNT, Wilfrid, *The Compleat Naturalist: A Life of Linnaeus*, London 1971.

BONAPARTE, Napoléon, *Correspondance générale*, I: *Les apprentissages 1784–1797*, [Paris] 2004.
—, *Correspondance générale*, II: *La campagne d'Égypte et l'avènement 1798–1799*, [Paris] 2005.

BOURRIENNE, Louis Antoine Fauvelet de, *Complete Memoirs of Napoleon Bonaparte*, ed. R.W. Phipps, London 1891.

BREYDENBACH, Bernhard von, *Peregrinatio in Terram Sanctam*, Mainz 1486.

BUHLE, Johann Gottlieb, *Literarischer Briefwechsel von Johann David Michaelis*, Leipzig 1794.

BUTLER, Arthur G., 'On the Lepidoptera collected in Socotra by Prof. I. B. Balfour', *Proceedings of the Zoological Society of London*, 1881, pp. 175–80.

CARDINALI, Sandro and Luigi PEPE, 'Gaspard Monge e la Spedizione in Egitto', *I Castelli di Yale*, 4, 1999, pp. 107–44.

CHARDIN, John, *Journal du voyage du Chevalier Chardin en Perse et aux Indes Orientales, par la Mer Noire et par la Colchide*, London 1686.

CHARLES-ROUX, François, *Les origines de l'expédition d'Égypte*, Paris 1910.

CROSLAND, Maurice, *The Society of Arcueil: A View of French Science at the Time of Napoleon I*, Cambridge, Mass. 1967.

CROSSLAND, Cyril, *On Forsskål's Collection of Corals in the Zoological Museum of Copenhagen*

(Spolia Zoologica Musei Hauniensis I), Copenhagen 1941.

DARWIN, Charles, *Narrative of the Surveying Voyages of His Majesty's Ships Adventure and Beagle between the Years 1826 and 1836, Describing their Examination of the Southern Shores of South America, and the Beagle's Circumnavigation of the Globe. Journey and Remarks 1832–1836*, London 1832–6.
— *On the Origin of Species by Means of Natural Selection, or the Preservation of Favoured Races in the Struggle for Life*, London 1859.

DEELMAN, J.C., 'The Life and Times of Commandeur Déodat De Dolomieu (1750–1801)' *Low-temperature Formation of Dolomite and Magnesite*, Compact Disc Publications, Eindhoven 2003, pp. 486–505.

DENON, Vivant, *Voyages dans la Basse et la Haute Égypte, pendant les campagnes de Bonaparte, en 1798 et 1799*, 2 vols., Paris 1802.

DHOMBRES, Nicole, 'Napoléon et les scientifiques, 1:1779–1798', *Revue du Souvenir Napoléonien*, 430, 1985, pp. 2–26.

DIERKENS, Alain and Jean-Marie DUVOSQUEL, eds., *Henri-Joseph Redouté et l'expédition de Bonaparte en Égypte*, Brussels 1993.

EHRENBERG, Christian Gottfried, *Naturgeschichtliche Reisen durch Nord-Afrika und West-Asien 1820 bis 1825 von W.F. Hemprich und C.G. Ehrenberg*, Berlin 1828.
— *Symbolae physicae seu Icones et descriptiones corporum naturalium novorum aut minus cognitorum quae ex itineribus Libyam Aegyptum Nubiam Dongalam Syriam Arabiam et Habessiniam ...*, Berlin 1828–99.
— 'Neue Beobachtungen uber blutartige Erscheinungen in Aegypten, Arabien und Sibirien', *Annalen der Physik und Chemie*, 8, 1830, pp. 477–514.

FERGUSON, Walter W., *The Mammals of Israel*, ed. Susan Menache, Jerusalem and New York 2002.

FORSSKÅL, Pehr, *Descriptiones animalium: avium, amphibiorum, piscium, insectorum, vermium; quae in intinere Orientali observavit Petrus Forskål. Post mortem auctoris edidit C. Niebuhr. Adjuncta est Materia Medica Kahirina atque tabula Maris Rubri geographica*, Copenhagen 1775.
— *Flora Aegyptiaco-Arabica sive Descriptiones plantarum, quas per Aegyptum Inferiorem et Arabiam Felicem detexit, illustravit Petrus Forskål. Post mortem auctoris editit Carsten Niebuhr. Accedit tabula Arabiae Felicis geographico-botanica*, Copenhagen 1775.
— *Icones rerum naturalium: quas in itinere Orientali depingi curavit Petrus Forskål. Edidit Carsten Niebuhr*, Copenhagen 1776.
— *Resa till Lycklige Arabien: Petrus Forsskåls Dagbok 1761–1763*, ed. with introduction by Arvid Hj. Uggla, Uppsala 1950.

FRIES, Th. M., J. M. HULTH and A.Hj. UGGLA, *Bref och skrifvelser af och till Carl von Linné. Första afdelningen*, vol. VI, *Bref till och från Svenska enskilda personer*, Stockholm 1912.

FRIIS, Ib, 'De botaniske resulater af Den Arabiske Rejse', in *Den Arabiske Rejse 1761–1767*, ed. Stig Rasmussen, Copenhagen 1990, pp. 117–28.
— 'Proposals to Amend the International Code of Botanical Nomenclature [ICBN] and Report of the Special Committee on Binary Combinations', *TAXON*, 42, 1992, pp. 343–50.

FÜCK, Johann, *Die arabischen Studien in Europa bis in den Anfang des 20. Jahrhunderts*, Leipzig 1955.

GEOFFROY SAINT-HILAIRE, Étienne, *Philosophie anatomique des organes respiratoires*, Paris 1818.
— *Principes de philosophie zoologique*, Paris 1830.
— *Lettres écrites d'Égypte: a Cuvier, Jussieu, Lacépède, Monge, Desgenettes, Redouté Jeune, Norry, etc., aux professeurs du Muséum et a sa famille*, ed. E.-T. Hamy, Paris 1901

GEOFFROY SAINT-HILAIRE, Étienne, Jules-César SAVIGNY *et al.*, *Description de l'Égypte, ou Recueil des observations et des recherches qui ont été faites en Égypte pendant l'expédition de l'armée française: Histoire naturelle*, 2 vols., Paris 1809–12.
GILLISPIE, Charles Coulston, 'Scientific Aspects of the French Egyptian Expedition

1798–1801', *Proceedings of the American Philosophical Society*, 133, 1989, pp. 447–74.

GRIFFTHS, Mark, 'Clifford's Banana: How Natural History was Made in a Garden', in *The Linnaean Collections*, ed. Mary J. Morris and Leonie Berwick, London 2007, pp. 19–36.

GROVES, Colin, 'Notes on Gazelles: IV. The Arabian Gazelles Collected by Hemprich and Ehrenberg', *Zeitschrift für Saugetierkunde*, 48, 1983, pp. 371–81.
— 'The Taxonomy of Arabian Gazelles', in *The Gazelles of Arabia* , ed. Kushal Habibi, Abdulaziz H. Abuzinada and Iyad A. Nader, Riyadh 1997, pp. 24–51.

HAMMERSTEIN, Notker, 'The Enlightenment', in Hilde de Ridder-Symoens, ed., *Universities in Early Modern Europe (1500–1800) (A History of the University in Europe*, vol. 2), Cambridge 1996, pp. 621–40.

HAMILTON, Alastair, *Europe and the Arab World: Five Centuries of Books by European Scholars and Travellers from the Libraries of the Arcadian Group*, Dublin, London and Oxford 1994.
— *The European Legacy: Arab and Islamic Culture in the Heritage Library of Qatar Foundation*, Doha 2006.
— *The Copts and the West 1439–1822: The European Discovery of the Egyptian Church*, Oxford 2006.

HAMILTON, Alastair and Francis RICHARD, *André du Ryer and Oriental Studies in Seventeenth-Century France*, Oxford 2004.

HANSEN, Thorkild, *Det Lykkelige Arabien: en dansk ekspedition 1761–67*, Copenhagen 1962
— *Arabia Felix: The Danish Expedition of 1761–1767*, London 1964.

HASSELQUIST, Frederick, *Iter Palaestinum eller Resa til Heliga Landet, förrätta ifrån år 1749 til 1752, med beskrifningar, rön, anmärkningar, öfver de märkvärdigaste naturalier, på hennes Kongl. Maj:ts befallning. Uitgifven af Carl Linnaeus*, Stockholm 1757.
— *Voyages and Travels in the Levant; In the Years 1749, 50, 51, 52. Containing Observations in Natural History, Physick, Agriculture, and Commerce: Particularly on the Holy Land, and the Natural History of the Scriptures*, London 1766.

HENZE, Dietmar, 'Carsten Niebuhrs Bedeutung für die erdkunde von Arabien', preface to Carsten Niebuhr, *Beschreibung von Arabien*, Graz 1969 (reprint of the 1772 Copenhagen edition).

HEPPER, F.N. and I. FRIIS, *The Plants of Pehr Forsskål's 'Flora Aegyptiaco-Arabica'*, Kew 1994.

HERODOTUS, *History*, transl. A.D. Godley, The Loeb Classical Library, 4 vols., London 1960.

HEROLD, Christopher, *Bonaparte in Egypt*, London 1963.

HERON, Robert, *Travels through Arabia, and Other Countries in the East, Performed by M. Niebuhr*, 2 vols., Edinburgh 1792.

HOGARTH, David George, *The Penetration of Arabia: A Record of the Development of Western Knowledge Concerning the Arabian Peninsula*, London 1904.

HÜBNER, Ulrich, 'Johann David Michaelis und die Arabien-Expedition 1761–1767', in Josef Wiesehöfer and Stephan Conermann, *Carsten Niebuhr (1733–1815) und seine Zeit*, Stuttgart 2002.

KARIG, Joachim S., 'A Prussian Expedition to Egypt in 1820: Heinrich von Minutoli', in *Travellers in Egypt* , ed. Paul Starkey and Janet Starkey, London 2001.

KECK, Egon, 'De hebraiske manuskripter og Kennicott-bibilen', in Stig Rasmussen, *Den Arabiske Rejse 1761–1767: En dansk ekspedition set i videnskabshistorisk perspektiv*, Copenhagen 1990.

KLAUSEWITZ, Wolfgang, 'The Red Sea Explorers Wilhelm Hemprich, Christian Ehrenberg and Eduard Rüppell', *Zoology in the Middle East*, 27, 2002, pp. 7–12.

KLAUSEWITZ, W. and J.G. NIELSEN, 'On Forsskål's Collection of Fishes in the Zoological Museum of Copenhagen', *Spolia*

Zoologica Musei Hauniensis, 22, 1965, pp. 1–29.

KLAVER, J. M. I., *Geology and Religious Sentiment: The Effect of Geological Discoveries on English Society and Literature Between 1829 and 1859*, Leiden, New York and Cologne 1997.

LACROIX, Alfred, *Déodat Dolomieu. Membre de l'Institut National (1750–1801). Sa vie aventureuse, sa captivité, ses oeuvres, sa correspondence*, 2 vols., Paris 1921.

LAISSUS, Yves, *L'Égypte, une aventure savante avec Bonaparte, Kléber, Menou 1798–1801*, Paris 1998.

LAURENS, Henry, *Les origines intellectuelles de l'expédition d'Égypte: l'orientalisme Islamisant en France (1698–1798)*, Istanbul and Paris 1987.

LECROY, Mary, *Type Specimens of Birds in the American Museum of Natural History*, New York 2003.

LICHTENSTEIN, M. H. C., *Verzeichnis der Doubletten des zoologischen Museums der Königliche Universität zu Berlin nebst Beschreibung vieler bisher unbekannter Arten von Säugethieren, Vögeln, Amphibien und Fischen*, Berlin 1823.
— *Darstellung neuer oder wenig bekannter Säugethiere*, Berlin 1827.

LINNAEUS, Carolus, *Systema naturae*, Vienna 1767.

LYELL, Charles, *Principles of Geology: Being an Attempt to Explain the Former Changes of the Earth's Surface, by Reference to Causes now in Operation*, 3 vols., London 1830–3.

MAEMPEL, George Zammit, 'The Arabian Voyage of 1761–67 and Malta: Forsskål and his Contribution to the Study of Local Natural History', *Proceedings of History Week*, 1992, pp. 35–76.

MALRAUX, André, *Vie de Napoléon par lui-même*, [Paris] 1930.

MARAS, Raymond J., 'Napoleon: Patron of Science', *The Historian*, 11, 1958, pp. 46–62.

MARSOT, Afaf Lutfi al-sayyid, *Egypt in the Reign of Muhammad Ali*, Cambridge 1984.

MICHAELIS, Johann David, *Fragen an eine Gesellschaft gelehrter Männer, die auf Befehl Ihro Majestät des Königes von Dänemark nach Arabien reisen*, Frankfurt 1762.
— *Literarischer Briefwechsel*, ed. Joh. Gottlieb Buhle, Leipzig 1794.

NECTOUX, H[ypolite], *Voyage dans la Haute Égypte, au dessus des cataractes: avec des observations sur les diverses espèces de séné qui sont répandues dans la commerce*, Paris 1808.

NIEBUHR, Carsten, *Beschreibung von Arabien: Aus eigenen Beobachtungen und im Lande selbst gesammleten Nachrichten abgefasset*, 2 vols., Copenhagen 1772.
— *Reisebeschreibung nach Arabien und andern umliegenden Ländern*, 3 vols., Copenhagen 1774, 1778, 1837.

NIEBUHR, Barthold Georg, 'Carsten Niebuhrs Leben', in Carsten Niebuhr, *Reisebeschreibung nach Arabien und andern umliegenden Ländern*, ed. Wolfgang Stammler, Zürich 1992.

O'NEILL, J. C., *The Bible's Authority: A Gallery of Thinkers from Lessing to Bultmann*, Edinburgh 1991.

PALLARY, Paul, *Marie-Jules-César Savigny, sa vie et son oeuvre*, Cairo 1931–4.

PITTON DE TOURNEFORT, Joseph, *Relation d'un voyage du Levant*, 2 vols., Paris 1717.

PLINY, *Natural History*, transl. H. Rackham, The Loeb Classical Library, 10 vols., London 1968.

POCOCKE, Richard, *A Description of the East and Some Other Countries*, 2 vols., London 1743–5.
— *Voyage de Richard Pockocke en Orient, dans l'Égypte, l'Arabie, la Palestine, la Syrie, la Grèce, la Trace, &c. &c. &c.*, 7 vols., Paris 1772.

PRATT, Mary Louise, *Imperial Eyes: Travel Writing and Transculturation*, London 1992.

PROVENÇAL, Philippe, 'The Cultural

Significance of the Results in Natural History Made by the Arabic Travel 1761–1767.
— Peter Forsskål's Contribution to Arabic Lexicography', in *Carsten Niebuhr (1733–1815) und seine Zeit*, ed. Josef Wiesehöfer and Stephan Conermann, Stuttgart 2002.

RASMUSSEN, Stig, ed., *Den Arabiske Rejse 1761–1767: En dansk ekspedition set i videnskabshistorisk perspektiv*, Copenhagen 1990.
— 'Frederick Christian von Haven og de filologiske resulater', in *Den Arabiske Rejse 1761–1767*, ed. Stig Rasmussen, Copenhagen 1990, pp. 303–37.
— 'Die arabische Reise', preface to Carsten Niebuhr, *Reisebeschreibung nach Arabien und andern umliegenden Ländern*, ed. Wolfgang Stammler, Zürich 1992.
— '"Niebuhriana" in Kopenhagen', in Josef Wiesehöfer and Stephan Conermann, eds., *Carsten Niebuhr (1733–1815) und seine Zeit*, Stuttgart 2002, pp. 43–46.

RAUWOLFF, Leonhart, *Aigentliche Beschreibung der Raiß so er vor dieser zeit gegen Auffgang inn die Morgenländer fürnemlich Syriam, Iudaeam, Arabiam, Mesopotamiam, Babyloniam, Assyriam, Armeniam etc. nicht ohne geringe mühe und große gefahr selbs volbracht*, Lauingen 1582–3.

RAY, John, *A Collection of Curious Travels and Voyages*, London 1738.

RIEBECK, Emil, 'Tagebuch einer Reise von Kairo nach den Gebirgen am Rothen Meer, *Mittheilungen des Vereins für Erdkunde zu Halle a.S.*, 1881, pp. 40–50.

RIMBAUD, Arthur, *Les lettres manuscrites de Rimbaud*, ed. Claude Jeancolas, Paris 1997.

SAID, Edward W., *Orientalism* [1st edn., New York 1978], Harmondsworth 2003.
— *Culture and Imperialism*, London 1994 .

SALT, Henry, *A Voyage to Abyssinia, and Travels into the Interior of that Country, Executed under the Orders of the British Government, in the Years 1809 and 1810: In Which Are Included, An Account of the Portuguese Settlements on the East Coast of Africa, Visited in the Course of the Voyage; A Concise Narrative of Late Events in Arabia Felix*, Philadelphia and Boston 1816.

SAUNERON, Serge, ed., *Le voyage en Égypte de Pierre Belon du Mans 1547*, Cairo 1970.

SCHWEINFURTH, Georg, *Ein Besuch auf Socotra mit der Riebeck'schen Expedition: Vortrag in der zweiten oeffentlichen Sitzung der 56. Versammlung Deutscher Naturforscher und Aerzte zu Freiburg I.B. 1883*, Freiburg i.B. 1884.
— *Auf unbetretenen Wegen in Aegypten: Aus eigenen verschollenen Abhandlungen und Aufzeichnungen*, Hamburg and Berlin 1922.

SCLATER, P.L. and G. HARTLAUB, 'On the Birds Collected in Socotra by Prof. I.B. Balfour', *Proceedings of the Zoological Society of London*, 1881, pp. 165–75.

SHAW, Thomas, *Travels, or Observations Relating to Several Parts of Barbary and the Levant*, London 1738.

SHEENAN, Jonathan, *The Enlightenment Bible: Translation, Scholarship, Culture*, Princeton and Oxford 2005.

STRESEMANN, Erwin, *Hemprich und Ehrenberg: Reisen zweier naturforschender Freunde im Orient geschildert in ihren Briefen aus den Jahren 1819–1826*, Berlin 1954.

TALLEYRAND, Charles Maurice de, *Essai sur les avantages à retirer de colonies nouvelles dans les circonstances présentes*, Paris 1797.

TASCHENBERG, Otto, 'Beiträge zur Fauna der Insel Sokotra, vorzüglich nach dem Hernn Dr. Emil Riebeck aus Halle a.S. gesammelten Materiale zusammengestellt', *Zeitschrift für Naturwissenschaften*, 1883, pp. 157–85.

THÉVENOT, Jean de, *Relation d'un voyage fait au Levant*, Paris 1664.
— *Suite du voyage de Levant*, Paris 1674.

TRIEDER, Simone, *Emil und Paul Riebeck: Söhne des Grossindustriellen Carl Adolph Riebeck*, Halle 2008.

W., M. v.d., 'Dr. E. Riebeck's Expedition', *Hallesches Tageblatt*, 28 December 1880.
WARNER, Nicholas, *The True Description*

of Cairo : A Sixteenth-Century Venetian View, 2 vols., Oxford 2006.

WELLSTED, J. R., *Travels in Arabia*, London 1838.

WIESEHÖFER, Josef and Stephan CONERMANN, eds., *Carsten Niebuhr (1733–1815) und seine Zeit*, Stuttgart 2002.

WOLFF, Torben, 'De zoologiske resulater af Den Arabiske Rejse', in *Den Arabiske Rejse 1761–1767*, ed. Stig Rasmussen, Copenhagen 1990, pp. 231–7.

YARON, Isaac, Tom Schiøtte and Gudrun Wium-Andersen, 'A Review of Molluscan Taxa Described by P. Forsskål and C. Niebuhr with Citation of Original Descriptions, Discussion of Type-Material Available and Selection of Some Lectotypes', *Steenstrupia*, 12, 1986, pp. 157–200.

INDEX